International Oilfield Surface Fac
Analysis for Electrical Design

Kun Ma · Yewei Mei · Xiaolong Meng ·
Zhaoxia Liu · Jingjun Huang · Zhiwei Liu

International Oilfield Surface Facilities: Safety Analysis for Electrical Design

Kun Ma
Department of Electrical
Beijing Company of China Petroleum
Engineering and Construction Corporation
Beijing, China

Yewei Mei
Department of Electrical
Beijing Company of China Petroleum
Engineering and Construction Corporation
Beijing, China

Xiaolong Meng
Department of Electrical
Beijing Company of China Petroleum
Engineering and Construction Corporation
Beijing, China

Zhaoxia Liu
Department of Electrical
Beijing Company of China Petroleum
Engineering and Construction Corporation
Beijing, China

Jingjun Huang
Department of Electrical
Beijing Company of China Petroleum
Engineering and Construction Corporation
Beijing, China

Zhiwei Liu
Department of Electrical
Beijing Company of China Petroleum
Engineering and Construction Corporation
Beijing, China

ISBN 978-981-16-3106-1   ISBN 978-981-16-3104-7   (eBook)
https://doi.org/10.1007/978-981-16-3104-7

Jointly published with Petroleum Industry Press
The print edition is not for sale in China (Mainland). Customers from China (Mainland) please order the
print book from: Petroleum Industry Press.

This Springer imprint is published by the registered company Springer Nature Singapore Pte Ltd.
The registered company address is: 152 Beach Road, #21-01/04 Gateway East, Singapore 189721,
Singapore

Preface

As the focal point of oilfield surface facilities engineering, the essential safety of electrical design is not only the necessary condition to ensure the normal production of all kinds of oilfield equipment, but also the key factor to prevent accident and protect personal safety from the original source. With the international oil field market opening to the world step by step, the domestic specifications of oilfield electrical design are gradually in line with the international standards, and the relevant safety management is gradually approaching to the international higher-class level, and the design essential safety concept is also improved consequently.

In this book, it describes the safety concerned topics together with technical analysis in the electrical engineering of international oil field surface facilities, including Electrical System Safety and Operational Risk Analysis (ELSOR), power supply & distribution design and safety analysis, overhead transmission line design and hazardous area classification for surface facilities. It is with the actual engineering application as the principle line in the following chapters, and to provide reference for international oilfield engineering concisely and practically.

ELSOR analysis illustrates the details of oilfield electrical project technical comprehensive reviewing from safety and operability risk point of view in order to provide the basis for accident prevention measures and emergency plan, which includes identifying potential risk, analysis and assessment, advice, and solutions in the electrical system design and skid engineering.

Power supply and distribution design safety analysis describes different understandings in load calculation, overvoltage, safety clearance, fire protection of substation in different codes, it also illustrates different specification of phase color, battery, electric leakage protection, and electrical leakage protection and electrical safety protection in different standards with reference to actual engineering matters through technical comparison. It analyzes and summarizes the key safety points in power supply and distribution design of international oil field surface engineering, and could be as the guideline of related design safety issues.

Overhead transmission line design safety analysis of international oil field surface engineering focus on mainly the safety coefficient selection, insulation coordination, and lightning protection and safety protection of overhead transmission lines.

Hazardous area classification plays the key role in electrical safety of international oil field surface engineering, which provides the fundamental basis for the overall layout of oil and gas process stations, and provides qualitative analysis and quantitative evaluation for the hazard and operability study (HAZOP) of the project. This section conducts benchmarking analysis on the current popular explosion-proof analysis guidance such as APIRP500/505, IP15, IEC60079, SY/T6671 and other domestic and foreign standards, and illustrates the difference between standards in actual hazardous division of the same equipment, which could be as the reference and guidance in electrical design safety of international oil fields.

Beijing, China   Kun Ma
December 2020

Contents

Chapter 1
Electrical System Safety and Operability Review (ELSOR) for International Oilfield Surface Facilities

1 Overview

The design safety philosophy of electrical system in international oil field surface engineering is integrated with advanced preview control and comprehensive treatment. In oil field surface facilities, electrical system is the heart to ensure reliable production, the detection of potential risk to ensure safe production becoming the primary task, and the potential risk analysis should be based on risk determination logic, applying scientific and rigorous method to evaluate the safety of engineering, construction and operation. Safety in electrical systems of international oilfield refers to the safety of personnel and equipment under normal, emergency and maintenance conditions, and equipment safety link to personnel safety directly. Therefore, safety judgment from the view of personnel and equipment is the key points of electrical system safety analysis in international oil field.

Electrical System Safety and Operability Review (ELSOR) is in accordance to scientific procedures and methods, from the point of power system view to identify, analyze and evaluate potential risk of project or packaged equipment engineering. Provide improvement suggestions based on identified risks. Evaluate the safety and operability of the electrical system; provide the basis for accident prevention measures and emergency plans.

The main purpose of ELSOR is to review the safety and operability of electrical system and equipment, and ELSOR shall be conducted by engineers from construction, production, maintenance, design, supervision and manufacture together. ELSOR includes a rigorous analysis of the safety, integrity, reliability and operability of the design documents to ensure that the project achieves its intended design objectives in a real construction scenario. During analysis, experts of all specialties collectively identify potential deviations from the design purpose, analyze its possible causes and evaluate its corresponding consequences. ELSOR is adopted in the analysis process, combining with relevant electrical parameters, etc., to carry

© Petroleum Industry Press 2022
K. Ma et al., *International Oilfield Surface Facilities: Safety Analysis for Electrical Design*, https://doi.org/10.1007/978-981-16-3104-7_1

out from partial analysis to overall analysis according to different systems, and identify possible problems, causes, consequences and measures that should be taken in normal or abnormal situations.

ELSOR has established a collective assessment method for electrical system design safety. All engineering parties participate in the review collectively, and make a complete analysis of the electrical system through brainstorming, so as to avoid one-sidedness. ELSOR assessment is the review of system—unit—equipment. The overall review process is relatively comprehensive, which is conducive to the discovery of various potential risks of system, equipment, operation and maintenance. ELSOR assessment has a complete logical structure to grasp easily.

2   ELSOR Review Guidelines

ELSOR is a process to assess the electrical system and equipment during design stages or existing operating plants to verify the design and to identify design inadequacies that would lead to potential failures during installation and operation leading to loss of power supply, equipment failures and unsafe conditions to human beings. ELSOR provides a systematic review methodology to address any discrepancies and/or inadequacies and non-compliance of standards, codes of practices, statutory regulations and project design basis in design parameters, philosophy as well as installation, operation and maintenance of electrical-related system and equipment. Recommendations will be provided focusing on the system design, layout, sizing design, installation, and operation, and maintenance related activities and procedures.

There are two assessment aspects involved in ELSOR which are described below:

(1)   Electrical System Safety & Integrity Design (ESSID)

- Assess the integrity and operability of electrical system and equipment for safe and reliable operation of the plant.
- Provide recommendation for modifications to system, layout and design based on integrity, operability and safety issues.

(2)   Electrical System Task & Operational Safety (ESTOS)

- Assess the potential electrical hazards due to human activities during normal, emergency and maintenance operations scenarios.
- Provide recommendation for modifications in design, installation, operation and maintenance procedures to overcome integrity, operability and safety issues.

2.1 *ELSOR Objectives*

Assess the integrity and operability of electrical system and equipment for safe operation of the plant using ELSOR guidelines in general.

Assess the potential electrical hazards due to human activities and errors using Electrical System Task and Operational Safety (ESTOS).

Provide recommendations in the form of review report for any electrical systems modifications or improvements that are identified in the review process.

2.2 *ELSOR Scope of Review*

The assessment includes the review of the Single Line Diagrams, Schematic Diagrams, Electrical Equipment Layout and Hazardous Area Classifications (HAC), Calculations and Settings, Specifications including Design Basis and Equipment Specifications along with Data Sheets. Where similarities and duplications exist in the systems, it would suffice to assess only one set of such drawings/documents to avoid repetitive work.

2.3 *ELSOR Methodology*

The methodology or procedure adopted for the review is described as follows:

(1) To facilitate the examination, each system shall be divided into parts in such a way that the design intent for each part can be adequately defined. The drawings and documents to be reviewed under ELSOR shall be identified and listed. The identified drawings/documents shall be one part of the electrical system design that impacts the risk issues covered in ELSOR. It is not necessary to review those documents that don't play any role in the system operability, integrity and safety, such as cable schedule.

(2) The review shall be done in two stages of the project (i) at conceptual/basic design stage to assess the major issues at this point so as to allow them to be considered in the design and to facilitate future ELSOR studies. (ii) at detail design stage to carry out ELSOR when the design is almost frozen and only minimal changes are expected after the review. The review shall preferably be completed and implemented before the purchase order is placed.

(3) The Single Line Diagrams and other related drawings/documents shall be provided by the Project Design Team in soft and hard copies that have been duly reviewed and approved by competent authority for the design process and declared to be met for the ELSOR review requirement.

(4) The review shall be taken up in 'Bottom Up' approach. The 'Bottom Up' approach will start from documents relating to the lower-most level systems

and move up to the highest level in the hierarchy. This approach is used as systems are designed and sized based on what is being served downstream. For example, LV transformer sizing shall be reviewed first before the MV switchgear rating is assessed as the MV switchgear rating is based on the loads on the outgoing feeders of MV switchgear including the LV transformers.

(5) During the review, the design purpose within each part that is reviewed shall be described by the Design Team for the benefit of the review team members to understand the aspect which may be difficult to express in the drawings and documents.

2.3.1 ESSID

(1) ESSID shall be performed by "guide word" examination and deliberation for deviations from design intent based on guidelines given in this document to identify the causes and consequences leading to loss of integrity and operability with associated any safety issues and recommend a solution for implementation. All reasonable use and misuse conditions which are expected by the user should be identified. The guidelines given in this document cover reasonably most important aspects of design parameters that must be reviewed point by point as they directly impact the operability, integrity and reliability. However the guidelines given in this document are not exhaustive and there may be according to decisions based on experience and some other criteria that are different from those listed in the guidelines. Such decisions shall be deliberated among the review members in conjunction with the design team to achieve one technically acceptable consensus solution. It is preferable that such decisions are supported by any International Standards or Codes or any existing proven Practices. Special reference shall be made on such decisions in the review report.

(2) The deliberation shall be to examine elements of the electrical systems in terms of its characteristics/design intent either quantitatively or qualitatively using such analysis as below (Table 1).

(3) Design parameters of electrical systems, elements and components that are reviewed shall be recorded in the ESSID Worksheet indicating both compliance and non-compliance against each item and parameter reviewed.

2.3.2 ESTOS

(1) ESTOS is performed to assess the potential electrical hazards due to normal human activities including possible errors and to provide critical modifications that will ensure electrical system safety and also have improvements in reliability, integrity and operability in the installation, operation and maintenance of the electrical equipment and systems. All safety issues arising out of the review and the recommendation for improvement shall be listed in ESTOS worksheet. Assessment using ESTOS should be done immediately after ESSID

Table 1 Logical review key words

Analysis type	Definition
NO or NOT	Complete negation of the intent
More	Quantitative increase
Less	Quantitative decrease
As well as	Qualitative modification/increase
Part of	Qualitative modification/decrease
Reverse	Logical opposite of the design intent
Other than	Complete substitution
Early	Relative to clock time
Late	Relative to clock time
Before	Relative to order of sequence
After	Relative to order of sequence

has been completed, or else in parallel with ESTOS for the same selected part. The procedure is repeated for all identified parts of the system until the entire system is covered.

(2) Similar findings and recommendations is applied to other parts with similar equipment or system.

(3) The findings are recorded and issued as a report, for incorporations of recommendations, as given in the typical report appearing elsewhere in this document. ESTOS recommendations shall be accorded top priority for implementation as it relates to safety of personnel.

2.3.3 Flow Chart

The procedure is given below, in a flow chart form with sequence of events (Fig. 1).

3 ESSID Guidelines

For the purpose of ESSID review guidelines, the design documents will be divided into the following categories.

(1) Design Basis
(2) Distribution Transformer Sizing Calculation
(3) Distribution Transformer Specification and Datasheet
(4) LV Switchgears Specification and Datasheet
(5) Power Transformer Sizing Calculation
(6) Power Transformer Specification and Datasheet
(7) MV Switchgears Specification and Datasheet
(8) DC UPS Specification and Datasheet

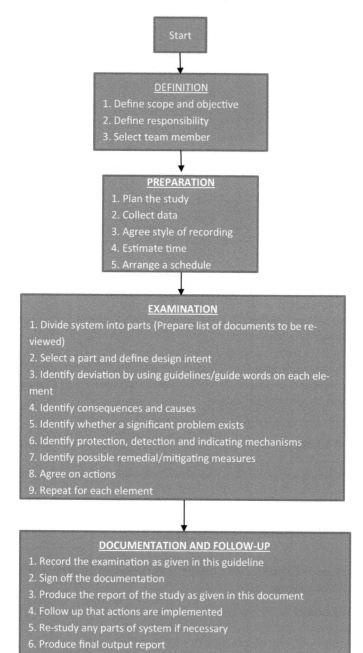

Fig. 1 ELSOR methodology block diagram

(9)   AC UPS Specification and Datasheet
(10)   Emergency Diesel Generator Specification and Datasheet
(11)   MV Motor Specification and datasheet
(12)   LV Motor Specification and Datasheet
(13)   MV Cable Specification and Datasheet
(14)   LV Cables Specification and Datasheet
(15)   Switchgear Layout
(16)   Battery Room Layout
(17)   Cable Routing Layout
(18)   Lighting Layout
(19)   Earthing Layout
(20)   LV DBs SLD
(21)   LV MCC/Switchgear SLD
(22)   MV Switchgear SLD
(23)   Motor Protections SLD
(24)   Distribution Transformer Protection SLD
(25)   Power Transformer Protection SLD
(26)   MV Switchgears Incomers Bus Couplers and Cable Feeders Protection
(27)   LV Switchgears Incomers Bus Couplers and Cable Feeders Protection
(28)   Protection Co-ordination and Relay Setting Calculations

General guidelines are given for each item in the following sections. However it should be emphasized here that these are standard based guidelines and may have to be reviewed based on specific project design and requirements. Any special designs shall be subjected to expert review and shall not be limited by these guidelines.

Order Priority:

(1)   Design Basis
(2)   Technical Specification
(3)   International Standards

In addition to the above, Local Statutory Requirements and Laws shall be abided by.

3.1 Design Basis

Design Basis shall be reviewed for the following,

(1)   Local Statutory Requirements. As local laws and regulations are specific for each project depending on the location of the project, involvement of Local agencies having full knowledge of local laws and regulations may be engaged for this purpose.
(2)   Violation/non-compliance of specified standards for the project. Care should be exercised in identifying a mandatory requirements and recommendations. Recommendations need not be considered as mandatory requirements.

However, if the project design needs such recommendations and is considered a critical factor, the same may be considered in the design.

(3) Safe limits of design parameters. This should include a generally well accepted design margins and safe design limits (such as Step and Touch Voltages).

(4) Safe interlocking requirements. Such interlocks shall be based on the level of automation and also the operation philosophy of the project.

(5) Earthing philosophy. This is importance because safety will be at risk when earthing is not implemented in strict adherence to safe earthing practices.

(6) Design Basis shall identify specific design margins, parameters, conditions (environmental) and minimum requirements. Generic statements such as 'adequate', 'enough margin', 'if required' etc. shall be avoided.

(7) Specific protection design requirement against exposed live parts shall be included in the Design Basis.

(8) Emergency situation related design requirements shall be clearly identified. This may include the emergency power requirement, autonomous times and minimum illumination level for egress and emergency lighting. All those services which have to support emergency situation shall be put into service automatically or shall be available for control from designated identified remote location to initial control based on emergency decisions. All abnormal conditions during emergency (such as non-availability of a critical equipment, overload or failure) shall be alarmed at control central location.

(9) Design ambient temperature shall be clearly stated for all types of equipment and for different installation conditions (such as indoor, outdoor etc.).

(10) Equipment Ingress Protection shall be stated for equipment installed in various environmental conditions (such as Indoor, outdoor etc.). Possibility of water ingress from sources such as leaking lines/flanges shall be reviewed thoroughly.

3.2   Distribution Transformer Sizing Calculation

(1) Continuous, Intermittent and Standby loads shall be clearly identified

(2) Continuous Loads shall be taken as a whole, except accounting for Load Factor (if deemed appropriate), in sizing calculation,

(3) Intermittent Loads shall be considered for sizing at one third of total Intermittent Loads

(4) Total Standby Loads shall be considered in the sizing and the largest standby load or 10% of the total standby loads whichever is largest shall also be added to the rating.

A safety margin of at least 10% shall be considered for the transformer rating. Safety margin is different from future loads provision. Future load provision should be based on the Design Basis or other documents (such as project contract specification) which may consider the future expansion provision requirements. In the absence of any such future expansion requirements a design safety margin of 10% shall be considered. If the transformers are rated 2 × 100%, the transformer rating of each transformer shall be adequate to cater to the total load.

3.3 Distribution Transformer Specification and Datasheet

(1) Transformer shall be specified for appropriate design ambient temperature as defined in the project design data. IEC 60076 indicates top oil temperature rise of 50 K and average winding temperature rise of 55 K overdesign ambient of 40 °C. If the project design ambient is different from this, appropriate changes in the allowable temperature rise shall be considered based on annual maximum, maximum monthly average and yearly average temperatures as guided in IEC 60076. In case the project specification for the electrical equipment clearly identifies design ambient and temperature rise, same shall be used in the specification. Design ambient will vary depending on the location (indoor and outdoor) and also provision of sun shade and forced ventilation etc.

(2) Rating of the transformer for both ONAN and ONAF mode of operation shall be clearly specified.

(3) Temperature rise for Oil and winding shall be clearly worded as 'Top Oil Temperature Rise' and 'Average Winding Temperature Rise'. These wordings are as used in the IEC standard and hence it would be advisable to follow the recommendations of the IEC Standard. Average oil temperature rise will be always lower than Top Oil Temperature Rise and hence care should be exercised in proper wording.

(4) Required OLTC/OCTC range and percentage voltage per step shall be clearly identified based on the Load Flow study.

(5) Tap changers current rating shall be at least 25% higher than the maximum current corresponding to the worst case operation at the lowest tap while delivering the rated power.

(6) Transformer through short circuit withstand capability shall be as recommended in the IEC for a fault level based on the switchgear feeding the transformer Fault Rating (not based on the switchgear short circuit level, based on short circuit study). Alternatively an infinite feeding source may be specified (at a higher cost) so as have higher safety margin.

(7) Transformer shall be specified to deliver the rated power at full tapping range.

(8) Percentage impedance shall be stated. Percentage impedance specified shall be same as that used in Short Circuit and Load Flow Studies including the tolerance and restrictions used in the studies. It shall be noted that higher impedance will reduce the short circuit level but increase the voltage drop and vice versa. Hence impedances specified shall be in strict accordance with the value used in short circuit and load flow studies.

(9)   Location of the transformer (indoor or outdoor) shall be clearly stated. Any unusual service conditions shall be specified.

(10)   Bushings shall be rated with at least 10% safety margin above the maximum current of the transformer corresponding to the worst case operation with lowest tap while delivering rated power.

(11)   Transformer shall be provided with pressure relief valve(s) and wired to trip the transformer instantly.

(12)   Transformer located indoor shall be dry type to mitigate fire possibility. Those oil filled transformers located in vulnerable areas that could jeopardize safety due to spread of fire, must be specified with high flammability (above 300 °C-designated with first letter 'K' per IEC standard) transformer insulating liquid.

3.4   LV Switchgears Specification and Datasheet

(1)   LV Switchgear short circuit rating shall be based on the short circuit study recommendations with a safety margin of about 10% and selected to the nearest standard rating according to IEC. The margin shall be not less than 5% as minimum. The short circuit calculated shall be that of half cycle short circuit value.

(2)   Short circuit study shall be calculated considering future expansion (and hence corresponding future capacity in the transformer) loads also and assuming at least 80% of the future additional capacity as lumped motor loads.

(3)   Short circuit study of switchgears having emergency generator, with online routine load testing by synchronizing with live switchgear, shall include emergency generator contribution also, as the testing process takes considerable time to complete and also the testing is frequent (tested every week or fortnight). Generators having their own dedicated load bank need not be considered in the short circuit study.

(4)   The rated peak make and withstand rating shall be based on in the Peak Current to symmetrical RMS short circuit ratio calculated in the short circuit study, subject to minimum of 1.5. The ratio shall refer to Table 2 [1].

Table 2 Values for the factor n (Peak/RMS)

R.m.s. value of short-circuit current kA	cos φ	n
$I \leq 5$	0.7	1.5
$5 < I \leq 10$	0.5	1.7
$10 < I \leq 20$	0.3	2
$20 < I \leq 50$	0.25	2.1
$50 < I$	0.2	2.2

(5)   The switchgear bus rating shall be higher, by a safety margin of about 10%, than the maximum transformer current corresponding to worst case operation while delivering rated power.

(6)   Design ambient temperature shall be according to the design basis or as given in the project Data.

(7)   Ingress protection grade of switchgear shall be as followings:

   – Min. IP2X in Specialized distribution room.
   – Min. IP54 in harsh environment.
   – Min. IP23 in other places.

(8)   Where highly skilled personnel are expected to operate and maintain with proper Permit To Work (PTW) procedure following Lock-out and Tag-Out method, the switchgear with Form 4 separation may be adequate. But in case personnel with inadequate experience and training are expected to operate and maintain the switchgear Form 4b shall be specified.

(9)   Breakers racking-in and racking-out shall be possible only in breaker open position with the door closed.

(10)   Incomers shall be provided with surge arresters to protect from surges.

3.5 Power Transformer Sizing Calculation

(1)   Continuous, Intermittent and Standby loads shall be clearly identified,

(2)   Continuous Loads shall be taken as a whole, except accounting for Load Factor (if deemed appropriate), in the sizing calculation.

(3)   Intermittent Loads shall be considered for sizing at one third of total Intermittent Loads.

(4)   Total Standby Loads shall be considered in the sizing and the largest standby load or 10% of the total standby loads whichever is higher shall be added to the rating.

(5)   A safety margin of at least 10% shall be considered to arrive at the transformer rating. Safety margin is different from future loads provision. Future load provision should be based on the Design Basis or other documents (such as project contract specification) which may consider the future expansion provision requirements. In the absence of any such future expansion requirements a design safety margin of 10% shall be considered. If the transformers are rated $2 \times 100\%$, the transformer rating of each transformer shall be adequate to cater to the total load.

3.6 Power Transformer Specification and Datasheet

(1)   Transformer shall be specified for appropriate design ambient temperature as defined in the project design data. IEC 60076 indicates top oil temperature rise

of 50 K and average winding temperature rise of 55 K overdesign ambient of 40 °C. If the project design ambient is different from this, appropriate changes in the allowable temperature rise shall be considered based on annual maximum, maximum monthly average and yearly average temperatures as guided in IEC 60076. In case the project specification for the electrical equipment clearly identifies design ambient and temperature rise, same shall be used in the specification. Design ambient will vary depending on the location (indoor and outdoor) and also provision of sun shade and forced ventilation etc.

(2) Rating of the transformer in both ONAN and ONAF shall be clearly stated.

(3) Temperature rise for Oil and winding shall be clearly worded as 'Top Oil Temperature Rise' and 'Average Winding Temperature Rise'. These wordings are as used in the IEC standard and hence it would be advisable to follow the recommendations of the IEC Standard. Average oil temperature rise will be always lower than Top Oil Temperature Rise and hence care should be exercised in proper wording.

(4) Required OLTC/OCTC range and percentage voltage per step shall be clearly identified based on the Load Flow study.

(5) Tap changers current rating shall be at least 25% above the maximum current corresponding to the worst case operation with lowest tap operation while delivering the rated power.

(6) Transformer short circuit shall be as recommended in the IEC for a fault level based on the Fault rating (not based on switchgear short circuit level based on short circuit study) of upstream switchgear feeding the transformer

(7) Transformer shall be specified to deliver the rated power at full tapping range.

(8) Percentage impedance shall be stated. Percentage impedance shall be specified as used in Short Circuit and Load Flow Studies including any restriction in tolerance used in the studies.

(9) Location of the transformer (indoor or outdoor) shall be clearly stated.

(10) Bushings shall be rated for at least 10% above the maximum current of the transformer corresponding to the worst case operation with lowest tap operation while delivering rated power

(11) Transformers shall be provided with dry air breather along with air bag (diaphragm) for the conservator.

(12) Transformers shall be specified according to insulation co-ordination study subject to minimum requirement given in IEC 60076. In case no study is done a value higher of the two values given in IEC shall be selected to make it safer.

(13) Transformer shall be provided with the following safety features:

(a) Main tank pressure relief valve(s) with tripping
(b) Pressure rise sensor with tripping
(c) OLTC Pressure Relief Valve and tripping (for transformers rated 25 MVA and above)
(d) OLTC Oil surge relay
(e) Bucholtz alarm and trip.

3.7 MV Switchgears Specification and Datasheet

(1)   MV Switchgear short circuit rating shall be based on the short circuit study with a safety margin of not less than 10% and selected to the nearest standard rating according to IEC. The minimum safety margin shall be not less than 5%.

(2)   Short circuit study shall be calculated considering future expansion loads (and hence corresponding future capacity in the transformer) and assuming 100% of the future capacity as lumped motor load.

(3)   Short circuit study of switchgears having MV emergency generator with online routine load testing by synchronising with live switchgear shall include emergency generator contribution also as the testing process takes considerable time to complete and also testing is frequent (tested every week or fortnight). Generators having their own load bank need not be considered in the short circuit study.

(4)   The peak make and withstand rating shall be, as minimum, in the same ratio between Peak Current to symmetrical RMS short circuit current calculated in the short circuit study subject to minimum of 2.5 as recommended in the Standards. It shall be noted that this may be exceeded in some special cases where generators are connected directly to the switchgear without a step up transformer. Care should be exercised to have appropriate ratio specified in case if it exceeds 2.5.

(5)   DC component breaking capacity required for the breakers shall be checked from the short circuit study. If the DC component breaking capacity falls short for the breakers selected the manufacturer should be consulted.

(6)   The switchgear bus rating shall be higher by a safety margin of about 10% than the maximum transformer current corresponding to worst case operation while delivering rated power.

(7)   Design ambient temperature shall be according to the design basis or as given in the project Data.

(8)   Switchgear ingress protection shall be as identified in the Design Basis.

(9)   Switchgear incomers, outgoing feeders and bus sections shall be provided with surge arresters to protect against surges.

(10)   Switchgear shall be specified to have the bus bars fully sleeve insulated to avoid bus faults.

(11)   For personnel safety the following shall be specified.

       (a)   Interlocked earth switches shall be specified for incomers, bus coupler, outgoing feeders and bus sections for use in maintenance scenarios to protect personnel against unexpected live.

       (b)   Capacitive coupled live cable indicators shall be specified.

       (c)   Switchgear shall be specified to be arc resistant for a time duration equivalent to at least the backup protection operated breaker clearing time and the switchgear shall be provided with are exhaust flaps to let out arc products from every compartment of the switchgear including cable, breaker, bus compartments of every breaker and PT and earthing cubicles. Additionally the switchgear enclosure shall be specified to have minimum 2 mm thickness for ingress protection.

(d)   Breakers racking-in and racking-out shall be possible only in breaker open position with the door closed.

3.8   DC UPS Specification and Datasheet

(1)   DC UPS shall be specified for 2 × 100% rectifiers. The rating shall be on the basis of 100% load plus boost charging of completely drained batteries with a safety margin of 10%.

(2)   DC UPS batteries configuration shall be according to Design Basis. Where 2 × 50% design has been adopted provision shall be made for automatic connection of both batteries in parallel. Alternatively manual transfer shall be provided for operation during emergency.

(3)   Batteries shall be provided with ambient temperature compensation to adjust the battery charging current based on battery temperature. If this is not done the battery will fail prematurely.

(4)   Autonomous time shall be as defined in Design Basis.

3.9   AC UPS Specification and Datasheet

(1)   AC UPS rectifier shall be based on 2 × 100% rectifiers. The rectifier rating shall be on the basis of 100% load plus boost charging of completely drained batteries with a safety margin of 10%. The inverters shall be sized for 2 × 100% loads with a design margin of 10%.

(2)   AC UPS batteries configuration shall be according to Design Basis. Where 2 × 50% design has been adopted provision shall be made for automatic connection of both batteries in parallel. Alternatively manual transfer shall be provided for operation during emergency.

(3)   Batteries shall be provided with ambient temperature compensation to adjust the battery charging current based on battery temperature. If this is not done the battery will fail prematurely.

(4)   Autonomous time shall be as defined in Design Basis.

(5)   The uninterrupted change over shall be achieved within half cycle.

(6)   The short time rating shall be adequate to clear the fault in the largest rated feeder in 5 ms or less.

(7)   Incomers shall be provided with surge arresters.

(8)   Input/output isolation transformers shall be used.

3.10 Emergency Diesel Generator Specification and Datasheet

(1) Emergency Diesel generator rating shall be specified to be on continuous duty (S1) basis.

(2) Emergency generators shall be specified to be highly reliable and have the following

 – Dual Redundant AVR with automatic changeover
 – Dual Redundant starting batteries with automatic change over
 – Dual redundant starting motors

(3) Emergency generator shall be specified to have starting capability under extreme temperature variations as applicable for the project and hence where required lube oil pre heating shall be specified.

(4) To ensure positive starting pre lubrication shall be specified.

(5) Ingress protection of generator shall be specified one level higher than that required for the environment in which it is located. The minimum recommended ingress protection shall be IP32.

(6) Emergency generator shall be designed to be installed without any blocking of combustion air.

(7) Emergency generator shall be designed for Class F (H) insulation with Class B (F) temperature rise.

(8) Emergency generators shall be protected to trip only for over speed and electrical short circuit (if the emergency generator is critical for safety of plant and personnel as in offshore installations). Other protections shall only alarm.

3.11 MV Motor Specification and Datasheet

(1) MV motors shall be specified to have a margin of at least 10% hot locked rotor withstand time over the starting time. When this is not followed for economic reason locked rotor protection with speed sensor may be implemented.

(2) Motors shall be rated to have a safety margin of at least 10% above the driven equipment shaft input power requirement. Where the pump impeller is designed to be upgraded for a higher capacity the corresponding rating shall be used to select the motor.

(3) Motors shall be specified for Class F insulation with Class B temperature rise.

(4) Motors with rating above 20% of feeding transformer motor starting study shall be performed for successful acceleration and voltage dips. Soft starter shall be specified for the motor to limit the motor starting current if required.

3.12   LV Motor Specification and Datasheet

(1)   Motors shall be rated to have a safety margin of at least 10% above the driven equipment shaft input power requirement. When the pump impeller is designed to be upgraded for a higher capacity, the corresponding higher rating shall be used to select the motor.
(2)   Motors shall be specified for Class F insulation with Class B temperature rise.
(3)   Motors with rating above 20% of feeding transformer motor starting study shall be performed for successful acceleration and voltage dips. Soft starter shall be specified for the motor to limit the motor starting current if required.

3.13   MV Cable Specification and Datasheet

(1)   MV cables voltage ratings shall be selected based on the E/F clearing time as recommended in the IEC 60183. The voltage rating shall, in no case, be lower than maximum expected continuous operating voltage.
(2)   Short circuit current rating of cables shall be based on back up protection fault clearing time (relay operating time + 100 ms).
(3)   Cables protected by current limiting fuses may be sized for one cycle time delay and mechanical robustness.
(4)   Cables ratings shall be selected based on installation method and considered all de-rating factors according IEC 60364.
(5)   Fire rating of cables shall be according to project specification as defined in the Design Basis. As minimum, cables shall be specified to be fire retardant.

3.14   LV Cable Specification and Datasheet

(1)   LV cables shall be selected to have voltage rating of 0.6/1 kV.
(2)   Cables ratings shall be selected based on installation method and considered all de-rating factors according IEC 60364.
(3)   Fire rating of cables shall be according to project specification as defined in the Design Basis. As minimum, cables shall be specified to be fire retardant.

3.15   Switchgear Layout

(1)   33 kV switchgear shall be installed in dedicated room.
(2)   Switchgears shall be installed with following minimum clearances

   – The minimum clearances in front of the switchgears shall meet the requirements of NEC Section 110-34(a) and Table 110-34(a) as given below

- Switchgears having wall in the front: (i) 1250 mm for switchgears rated up to 3.3 kV (ii) 1600 mm for 6.6 kV and 11 kV switchgears (iii) 1800 mm for 33 kV switchgears.
- Switchgears facing each other: (i) 1500 mm for switchgears rated up to 3.3 kV (ii) 1800 mm for 6.6 and 11 kV switchgears (iii) 2700 mm for 33 kV switchgears.
- In addition to the above, switchgears clearances shall also be adequate to draw the breaker completely out from the switchgear and additionally have minimum 500 mm clearance to allow for handling and manoeuvring of breaker trolley by operator.
- Switchgears requiring rear access for cable connection or maintenance shall have minimum 900 mm, 1200 mm and 1500 mm clearances from the wall for 3.3 kV, 6.6 kV, 11 kV and 33 kV switchgears respectively. Switchgears that don't need rear access shall have a clearance of more than 300 mm.
- Switchgears roof clearance shall be according to manufacturer's recommendations and the requirements of other services such as cable trays, air ducts etc. shall also be considered. The requirements based on arc gas exhausts shall be strictly prescribed by the switchgear manufacturer.
- Clearance between two switchgears in the same row shall be minimum 1.5 m.

(3)   Switchgear rooms shall be provided with minimum two exits. Out of this one could be for equipment loading and unloading. The equipment door shall be suitable for moving in and out of the largest shipping dimension of the equipment located within the switchgear room.

3.16   Battery Room Layout

(1)   Batteries shall be located in well ventilated room with forced ventilation for batteries rated 20 kWh and above.

(2)   Battery room lighting fixtures shall be suitable for Zone 2 Ex IIC classified area.

(3)   Battery room lighting fixtures shall be installed on the walls (they shall not be installed above battery racks).

(4)   No power socket outlets shall be installed inside the battery room for portable/temporary loads. If required, they shall be installed outside the battery room.

(5)   Battery inter-cell connectors shall be insulated and the battery terminal posts shall be covered with insulated shrouds (may be allowed to have a small hole for voltage measurements during maintenance.

(6)   Batteries shall be installed above 0.8 m height (top of the battery) above ground.

(7)   Battery room shall be installed with emergency eye wash water equipment with a pressure switch for alarm.

(8)   Batteries shall be installed in rows and tiers giving minimum 800 mm between two adjacent trains.

(9)   Lighting control switch shall not be installed inside the battery room. If required, control switch shall be installed outside the room in an unclassified area.

(10)   No socket outlets shall be installed within the room.

(11)   Batteries charging shall be regulated based on room temperature to avoid overheating and premature failure of the batteries.

(12)   Capacity tests shall be done on batteries by actual loading to assess the health of the batteries at regular intervals as recommended by the manufacturer. Necessary provision shall be included in the design to test the batteries online.

(13)   Battery room floor shall be given a slope of 1:100 such that the acid will collect in one corner in case of acid spillage.

(14)   For flooded acid batteries the floor may be made acid proof.

3.17 Cable Routing Layout

(1)   Cables shall not be routed in the following areas

    (a)   Sources of heat

    (b)   Areas which are vulnerable for equipment failure leading to cables damage (due to release of steam, acid, hot liquids)

    (c)   Classified Zone 0 areas

(2)   Cables shall be well protected against damage as given below

    (a)   Below Road crossings (should be well protected in Hume pipes or culverts as required)

    (b)   Above road crossings (should have clear height to allow vehicles crossing)

    (c)   At points where vehicles movements are expected close to the installations

    (d)   At points close to flanged joints of pipes carrying hot fluids including steam and oil (shall be protected against impinging of hot fluid by suitable barrier)

    (e)   At points where objects may fall from above

    (f)   Cables that are meant to be used in hot environment (such as wiring on engines) shall be suitable for the temperatures to which they will be subjected to (mineral and Teflon cables)

(3)   Sharp bends shall be avoided. The bending radius shall be at least 10% more than that recommended by manufacturers. Cable tray bends shall be appropriate type and factory manufactured to avoid sharp edges damaging the cables (not site fabricated).

3.18 Lighting Layout

(1) One third of total lighting shall be emergency lighting.
(2) The minimum illumination levels along the exit routes shall meet the local regulations.
(3) Provision shall be available to test the emergency lighting.
(4) Transition lighting shall be used DC source or built in integral inverter lighting fixtures to provide lighting after loss of power until emergency generators comes into service.
(5) Lighting layout should be such that alternate fixtures are connected to different circuits.

3.19 Earthing Layout

(1) Earthing design shall achieve safe step and touch potential as recommended by IEEE80. The actual step and touch voltages shall below the tolerable step and touch voltages as given in Eqs. (29) and (32) of IEEE 80 standard as applicable to 50 kg body weight.
(2) Fence touch potential is often a critical area where the touch potential could exceed the limit despite having very low earth grid resistance. Care shall be exercised to add additional earthing rods and conductors to limit the touch potential within allowable limits. Fence earthing philosophy shall be clearly specified in the Design Basis. It is advisable to have the earth grid extent beyond the fence by at least 1.5 m and connect the fence to the earth grid at frequent intervals of about 10 m.
(3) Earth grid resistance shall be well below 1 Ω (below 0.5 Ω for major installations).
(4) Surge arresters shall be connected to a dedicated earth pits and routed in such a way that the connection is straight without loop and too many bends and is the shortest in length.
(5) System neutrals shall be earthed through earth conductor of adequate rating such that the conductor will not melt and break earth continuity for the expected fault current during earth fault. Earthing at two points is strongly recommended for high Voltage systems. The temperature attained during fault shall be at least 25% less than the melting temperature of the conductor material.
(6) Equipment earthing conductor shall be sized for maximum earth fault current and shall not melt at the maximum fault current losing earth connection integrity. All major equipment shall be earthed at two points with conductors each rated for full maximum earth fault current. Connections that are vulnerable for damage shall be mechanically protected.

3.20 LV DBs SLD

(1) Distribution boards fault level shall be correctly calculated without assuming arbitrary fault level. The MCBs fault ratings shall be adequate with a safety margin of at least 10%. Incomer breaker Fault rating may be selected to be same as that of the upstream feeding switchgear. Where the short circuit rating is high, adding incoming cable length can substantially bring down the fault level.

(2) Outgoing breakers/MCBs shall break the neutral also for circuits feeding to explosive areas.

DBs are often worked with live circuits and hence the DBs shall be specified with finger proof IP20 protection to avoid accidental contact with exposed live internal parts.

3.21 LV Switchgear/MCC SLD

(1) The Short Circuit rating of the switchgear shall be in line with short circuit study report and recommendations.

(2) Peak short circuit current rating shall be in line with the short circuit study report.

(3) Bus rating shall be based on infeed transformer maximum current for the worst scenario while delivering rated power with about 10% safety margin.

(4) Incoming cable/bus duct shall be designed for minimum current as given above in (3) with 10% minimum safety margin.

(5) Breakers shall be de-rated at the rate of 10% for every 5 °C increase in Design ambient above standard 40 °C (manufacturers ratings are for design ambient of 40 °C only and they need to be de-rated for use above this temperature).

3.22 MV Switchgear SLD

(1) The Short Circuit rating of the switchgear shall be in line with short circuit study report.

(2) Peak short circuit current rating shall be in line with the short circuit study report.

(3) Bus rating shall be based on infeed transformer maximum current for the worst scenario while delivering rated power with about 10% safety margin.

(4) Incoming cable/bus duct shall be designed for minimum current as given above in (3) with 10% minimum safety margin.

(5) Breakers shall be de-rated at the rate of 10% for every 5 °C increase in Design ambient above standard 40 °C.

(6)   Breaking current shall be less than 3A for breakers. This will limit the switching surges to safe levels.

3.23   MV Motor Protections SLD

(1)   Motors rated 1 MW and above shall be provided with following protections as minimum

   (a)   Instantaneous Over Current (50)
   (b)   Inverse time over current (51)
   (c)   Thermal Overload (49)
   (d)   Locked Rotor (51R)
   (e)   Unbalanced Loading (46)
   (f)   Under Voltage (27)
   (g)   Instantaneous Sensitive Earth fault (50G)
   (h)   Definite time Earth fault (51G)
   (i)   Motor Differential (87)
   (j)   Start Inhibit (Frequent Start) (66)
   (k)   Stator Temp (RTD)
   (l)   Bearing Temp (RTD)
   (m)   Bearing Vibration

(2)   Motors rated up to 1 MW shall be provided with following protections as minimum

   (a)   Instantaneous Over Current (50)
   (b)   Inverse time over current (51)
   (c)   Thermal Overload (49)
   (d)   Locked Rotor (51R)
   (e)   Unbalanced Loading (46)
   (f)   Under Voltage (27)
   (g)   Instantaneous Sensitive Earth fault (50G)
   (h)   Definite time Earth fault (51G)
   (i)   Stator Temp (RTD)
   (j)   Bearing Temp (RTD) (Motors rated 500 kW and above)
   (k)   Bearing Vibration (Motors rated 500 kW and above and vertical motors).

3.24   LV Distribution Transformer Protection SLD

(1)   LV Distribution transformers shall be provided with the following protections on the HV side of the transformer

   (a)   Inverse time over current (51)

- (b) Instantaneous Earth fault (50/50G)
- (c) Definite Time/Inverse Time Earth fault (51G)
- (d) Standby Earth Fault (500 kVA and above)
- (e) Thermal Overload (49) (Rated 2000 kVA and above)
- (f) Transformer Differential (87) (for transformers rated 2000 kVA and above)
- (g) Restricted Earth Fault (64REF) (for transformers rated 2500 kVA and above)
- (h) Winding Temperature Trip (for transformers rated above 1000 kVA)
- (i) Winding Temperature Alarm (for transformers rated above 1000 kVA)
- (j) Oil temperature Trip (for transformers rated above 1000 kVA)
- (k) Oil Temperature Alarm (for transformers rated above 1000 kVA)
- (l) Pressure relief Valve Trip
- (m) Bucholtz Alarm and Trip
- (n) Oil Pressure High Trip
- (i) Oil Level Low Alarm (transformer rated above 1000 kVA)
- (o) Instantaneous Over Current (50)

(2) Transformers shall be made to trip, instantly, all source breakers on operation of protection without any further delay for protections for which additional time delay cannot be afforded (e.g. Pressure Relief, Bucholtz).

3.25 Power Transformer Protection SLD

(1) Transformers shall be provided with the following protections on the HV side of the transformer

- (a) Instantaneous Over Current (50)
- (b) Inverse time over current (51)
- (c) Instantaneous Earth fault (50N)
- (d) Inverse Time Earth fault (51N)
- (e) Standby Earth Fault
- (f) Thermal Overload (49)
- (g) Transformer Differential (87)
- (h) Restricted Earth Fault (64REF)
- (i) Over Voltage (59) only for transformers exposed to public grid connection
- (j) Over fluxing (24) only for transformers exposed to public grid connection
- (k) Winding Temperature Trip
- (l) Winding Temperature Alarm
- (m) Oil temperature Trip
- (n) Oil Temperature Alarm
- (o) Pressure relief Valve Trip

- (p) Bucholtz Alarm and Trip
- (q) Oil Level Low and Low-Low Alarm
- (r) Main Tank Pressure HIGH Trip
- (s) OLTC Oil Surge Protection
- (t) OLTC Oil Pressure High Protection (For transformers Rated 25 MVA and above)
- (u) OLTC Pressure relief valve Protection

(2)   Transformers shall be made to trip all source breakers instantly on operation of protection without any further delay for those protections for which additional time delay cannot be afforded (e.g. Pressure Relief, Bucholtz).

3.26 MV Switchgears Incomers, Bus Coupler and Cable Feeder Protections

(1)   Incomers shall be provided with following protections

- (a) Inverse Time Overcurrent (51)
- (b) Definite time Overcurrent (51)
- (c) Inverse Time earth fault (51N) for solidly grounded systems
- (d) Inverse time earth fault (51G) for resistance grounded systems
- (e) Definite time earth fault (51N) for solidly grounded systems
- (f) Definite time earth fault (51G) for resistance grounded system
- (g) Time delayed UV (27)

(2)   Bus couplers shall be provided with following protections

- (a) Inverse Time Over Current (51)
- (b) Definite time Overcurrent (51)
- (c) Inverse Time earth fault (51N) for solidly grounded systems
- (d) Definite time earth fault (51N) for solidly grounded systems
- (e) Time delayed UV (27)

(3)   Cable feeders shall be provided with following protections

- (a) Inverse Time Over Current (51)
- (b) Definite time Overcurrent (51)
- (c) Inverse Time earth fault (51N) for solidly grounded systems
- (d) Inverse time earth fault (51G) for resistance grounded systems
- (e) Definite time earth fault (51N) for solidly grounded systems
- (f) Definite time earth fault (51G) for resistance grounded system
- (g) Time delayed UV (27).

3.27   LV Switchgears Incomers, Bus Coupler and Cable Feeder Protections

(1)   Incomers shall be provided with following protections

     (a)   Inverse Time Over Current (51)
     (b)   Definite time Overcurrent (51)
     (c)   Inverse Time earth fault (51N)
     (d)   Definite time earth fault (51N)
     (e)   Time delayed UV (27)

(2)   Bus couplers shall be provided with following protections

     (a)   Inverse Time Over Current (51)
     (b)   Definite time Overcurrent (51)
     (c)   Inverse Time earth fault (51N)
     (d)   Definite time earth fault (51N)
     (e)   Time delayed UV (27)

(3)   Cable feeders shall be provided with following protections

     (a)   Inverse Time Over Current (51)
     (b)   Definite time Overcurrent (51)
     (c)   Inverse Time earth fault (51N)
     (d)   Definite time earth fault (51N).

3.28   LV Diesel Generator Protections

(1)   Over Voltage (59) for generators rated 500 kW and above
(2)   Over Frequency (81) for generators rated 500 kW and above
(3)   Field winding insulation Failure (64R) for generators rated above 1 MW
(4)   Differential Protection (87) for generators rated 1 MW and above
(5)   Unbalanced Loading (46)
(6)   Thermal Overload (49)
(7)   Voltage Restraint Over Current (51 V) for generators rated 1 MW and above
(8)   Field failure (41) for generators rated 1 MW and above
(9)   Reverse Power (32) for generators rated for 500 kW and above
(10)   Phase over current (51) for generators rated less than 1 MW
(11)   Earth Fault (51N).

3.29   Protection Co-ordination and Relay Setting Calculations

Following shall be the guidelines for setting the relays in general. It may be noted that there may be specific requirement relating characteristics of particular make and type of relays. Manufacturer's relay manual must be consulted to get the full details as how the relays are to be set. The deciding factor for the settings shall be the Criteria given below in each case. It should be possible to fine tune the settings within these guidelines for better protections (lower operating times) and still maintain the discrimination. Plot of time current curves (TCC) is highly recommended as this will help visualize the complete range of power system operation and ensure selectivity of protections with adequate time discrimination.

(1) Relays are set in terms of CT secondary currents and hence the settings are to be converted into equivalent CT secondary currents. This is an area where mistakes happen often resulting in either unwanted spurious tripping or protection not operating when required, resulting in major failures.

(2) CT Ratios should be selected based on Maximum Demand with about 25% safety margin. The selected metering CTs shall have enough VA burden so as not to overload the CTs. CTs are not protected against over load and hence if overloaded above their rated burden CTs burst. The connected burdens shall be less than the CT rated burden by a margin of about 25%.

(3) CTs for protection shall be selected based on the vendors recommendations according to relay design and requirements. No two relay manufacturers have same requirement for same protection due to relay design, though there may be some similarities.

(4) MV Equipment shall use Class PX CTs for differential protections, while class 'P' CTs may be acceptable for LV equipment.

(5) Unit Schemes (such as Differential Protections) cover the power systems with instantaneous protection as primary protection. Other time delayed protections should be used as back up protections. If primary protection fails the backup protection should clear the fault and hence the protection scheme shall ensure there is always a backup protection with sufficient sensitivity, for every kind of short circuits and earth faults at all points in the system.

(6) Protections shall be set to be selective with enough time delay such that only bare minimum breakers to clear a fault will be tripped while leaving other healthy sections intact without any interruptions.

(7) The clearing time by back up protection shall not exceed the short time rating of the equipment being protected.

(8) Discrimination is achieved both by current and time settings. Current based discrimination is possible where there is substantial change in short circuit current level between the two protected points (For example primary and secondary sides of a transformer). Where current levels do not differ much discrimination is achieved by time grading.

(9) Following general rules shall be followed to achieve the discrimination between protections

   (a) The time discrimination between two protections located on the upstream and downstream of the network shall be 0.25 s in general, for the entire range of short circuit current (if possible) in the relevant circuit. The minimum discrimination time shall be 0.2 s. This recommendation is

based on numerical relays considering typical relay and CT accura-
cies and breaker fault clearing time. If, however, relay manufacturer
recommends different value same must be followed.

(b)   Generally time grading is done for normal operation of the system.
      However, depending on individual project requirement, grading may be
      considered for some other operating scenarios if desired (example one
      incomer feeding the switchgear with BC closed).

3.30   Relay Setting for MV Motors

(1)   Instantaneous Over Current (50): This shall be set at about 1.4 times the locked
      rotor current. Provide a small time delay of one cycle. Criteria shall be to set
      above the DC offset during starting.
(2)   Inverse time Over Current (51): This shall be set to pick up at above 10 to 30%
      of the FLC of the motor. The higher pick up is required as safety margin for
      possible short time overloads. If overload pattern is exactly known the pickup
      may be set with 10% margin over the short time overload.
(3)   Locked Rotor (51R): This shall be set to operate at about 10% more than the
      hot locked rotor withstand time.
(4)   Unbalanced Protection (46): This shall be set at about 25–30% of the FLC.
      Time delay may be set at 10 s.
(5)   Under Voltage (27): This may be set at about 70% with a time delay of 3 s. Exact
      setting shall be possible maximum under voltage during various operating
      scenarios of the system such as motor starting and motor stalling limit.
(6)   Motor Differential Protection (87): This may be set at about 10 to 15% for
      normal load and about 20–25% at and above 200% current (Follow relay manu-
      facturer's recommendations). CT errors, Relay error shall be the criteria for
      setting.
(7)   Start Inhibit (66): Follow relay setting guidelines by manufacturers Maximum
      only two hot starts and three cold starts are allowable in a sequence, with
      recommended time gap as given by motor manufacturer. Relay simulates the
      thermal image of the motor and inhibits starting when temperature exceeds the
      limit. Relay manufacturers setting guidelines shall be followed.
(8)   Instantaneous sensitive earth fault (50G): Set at about 25% with one cycle time
      delay. Setting should be more than the capacitive charging current of the cable.
(9)   Definite time earth fault (50G): Set at about 10% and 0.25 s delay. This is a
      backup protection to the above.

3.31   Relay Settings for Transformer

(1)   Instantaneous Over Current (50): The setting shall.

(a) be at least 130% of through fault current.
(b) be sensitive enough to operate at 75% of minimum fault current.
(c) be above the transformer inrush current to avoid tripping due to magnetic inrush current during energization.
(d) not interfere with all credible operating scenarios of loads connected to the transformer (such as largest standby motor starting over 100% standing base load and motors group re-acceleration).

(2) Inverse time over current (51): 51 should be set to pick up at about 125 to 150% of FLC. Higher margin is adopted to allow for short time overload (such as motor starting). Largest standby motor starting shall be plotted to check the setting is adequate.
(3) Instantaneous Earth fault (50N): Shall be set at about 50% with one cycle time delay.
(4) Definite time earth fault (51N): This shall be set at about 25% and additional 0.25 s time delay above the outgoing feeders.
(5) Differentia Protection (87): This shall be set for about 15% at normal current and about 25 to 40% at twice the rated current (Follow relay manufacturers recommendations). CT errors relay error and effect of OLTC on CT ratios shall be the considered in setting. Follow relay manufacturer's recommendations.
(6) Over Voltage protection (59): This shall be set to lie below the transformer over voltage withstand curve provided by transformer manufacturer. In the absence of this data this may set at 125% 3 s and 140% 0.5 s.
(7) Over fluxing protection (24): this shall be set to lie below the transformer V/f withstand curve provided by the transformer manufacturer. At least 10% safety margin shall be ensured for the relay setting plus breaker interruption time.
(8) The primary CTs on transformers with delta primary, will only sense 49% of the secondary earth fault current and hence the primary protection curve should lie below the transformer damage curves shifted lower by 49%.
(9) The phase protection curves shall lie below the mechanical damage curve of the transformer.

3.32 Relay Setting for Switchgears Incomers/Bus Couplers

(1) Definite time Over Current (51): Set at 300% with time delay of 0.25 s more than the outgoing feeders. The criteria shall be not to interfere with all credible operating scenarios including starting of largest standby motor on 100% standing base load and re-acceleration of group motors (if applicable). Curves shall be plotted to ensure this is achieved. Definite time should be sensitive enough to operate at about 75% of the minimum fault current.
(2) Inverse time over current (51): 51 should be set to pick up at about 125–150% of the transformer full load current. Higher setting is chosen to allow for any short time overload. If the over load is well known the setting may be done with 10% safety margin over the known overload.

(3)   Inverse time earth fault (51N): This shall be set at about 25%.
(4)   Definite time earth fault (51N): This shall be set at about 50% and shall be sensitive enough at 75% of minimum earth fault current.
(5)   Inverse time Protection of cable feeder (51): This shall be set to pick up at 125% of the maximum demand of the feeder.
(6)   Definite time earth fault of cable feeder (51N): This shall be set at 50%.

3.33 Relay Setting for Diesel Generator

(1)   Over Voltage (59): This shall be set to clear (set above) the transient over voltage conditions during load throw off and voltage recovery after clearance of a short circuit. If this data is not available it may be set at 125% 3 s and 140% 0.5 s. (The settings shall be approved or as recommended by the generator manufacturer.)
(2)   Over Fluxing (24): This shall be set to clear the generator withstand capability data provided the generator manufacturer with about 10% safety margin.
(3)   Field Winding Fault: Alarm at 100 k Ohms and trip at 25 k Ohms (follow manufacturer's recommendations).
(4)   Differential protection (87): Set at 10% for Normal current and 25% for current above 200%. The setting shall consider the CT errors and relay error (Follow relay manufacturer's recommendations).
(5)   Unbalance Loading (46): Alarm at 8% and trip at 10%.
(6)   Voltage Restraint Over Current (51 V): The pickup shall be set at about 125–150% and at 100% restraint the curve shall lie below the overload capability curve of generator.
(7)   Field failure (41): The impedance relay shall be set be offset and have its Centre equivalent to 2Xd" along the −X axis with an impedance circle of Xd/2 diameter. The time delay may be 4 to 5 cycles.
(8)   Reverse power (32R): Set at about 2% (Follow manufacturer's recommendation).
(9)   Phase Over Current (51): Set to pick up at 125% of FLC.
(10)   Inverse Time Earth fault (51N): Set to pick up at 25%.

4 ESTOS Review Guidance

ESTOS review shall focus on the safety aspects of the design. This shall include electrical shock due to potential rise during fault, contact with live electrical circuits, possible mechanical failure of the equipment with associated safety risks such as explosion and fire hazards. Equipment wise guidelines have been covered below for all credible failure modes. It must be noted that design can address inadequacies, however, it is uneconomical to over design to avoid failures 100% and hence it is impossible to avoid failures as they could also be caused by manufacturing defect, raw material quality and abuse of equipment during installation and subsequent operations.

4.1 Transformers

Transformers pose hazards in many ways. Chief among them are listed below

(1) Internal Failure—When this happens, lot of energy is released heating up the oil which thermally expands and may also produce arc and gas resulting in sudden increase in the internal pressure. To counter this, the main tanks shall be provided with pressure relief valves which will operate and release the oil to outside to reduce the pressure. It shall be noted that depending on the volume of oil and also for reliability, sometimes two relief valves will have to be provided for large transformers. The manufacturer shall be asked to provide calculation for adequacy of pressure relief valve capacity to contain the pressure rise within safe limits of the transformer tank and this should be included in the specification.

Also, when the relief valves operate the transformer shall be de-energized instantly without any time delay and hence it is advised to trip the breaker instantly by wiring the relief valve avoiding any intended or unintended delay. Failure of relief valve or delayed tripping might result in explosion of the main tank walls resulting in splashing of hot oil which will be dangerous to the people who might be in the proximity of the transformer during such explosions.

(2) In the above mentioned scenario, it is possible that the transformer catches fire in case of severe fault and (or) delayed opening of breaker. In such case, the oil and gas combination would fire and it may engulf the entire area and spread to the adjacent equipment and buildings. Spread of fire should be handled by suitable fire protection design. The usual designs adopted are (a) providing fire wall between transformers installed side by side and also making the building wall fire resistant. The protection requirement would depend on, how close the transformer is installed with respect to the buildings and adjacent transformers including the volume of oil used in the transformer. The requirements stated in NFPA 850 Standard shall be followed for the design.

(3) Since a fire break out is quite a possibility in transformers, despite providing means to prevent it, suitable fire suppression system should be considered for transformers with high velocity water spray by tapping water from plant fire water system.

(4) Similar design is required for OLTC section of the transformers also and a pressure relief valve and oil surge relay will be required to be installed (at least two protections should be installed).

(5) For transformers having the possibility of earth potential rise due to return of remote fault current earth grid design shall be adequate to limit the touch and

step voltages below the tolerable limits. The need for spreading gravel layer should be emphasized here in the design.

4.2   *Switchgears*

Switchgears have many hazards associated with them. Most of the Electrical faults within switchgears are not bolted type and often result in arcing. Arcing creates puffing action and building up pressure within the cubicle. Once cubicle enclosure fail, the hot plasma products of arc would spread causing serious damage to the personnel in the vicinity. Following are the identified hazards in the switchgear.

(1)   Arcing within the switchgear. Switchgear should be specified to be arc proof for a fault level equivalent to the switchgear rating and for a period of at least equivalent to the primary protection clearing the fault. The recommended time duration is 0.5 s. The switchgears should be assessed for arc energy level by conducting Arc Flash Studies and personal protective suites should be recommended based on NFPA 70E for all such operations which might result in arcing fault.

(2)   Switchgears shall be provided with interlock to prevent moving breakers (rack-in/rack-out) in closed position as this would result in flash-over.

(3)   Switchgears shall be provided with interlock between earthing switch and breaker such that (a) earthing switch can be closed only after operating breaker to isolated position (b) similarly breaker can be racked in only after earthing switch is open.

(4)   Wherever possible upstream feeding breaker earthing switch and the switchgear incomer earthing switches shall be interlocked such that (a) Upstream earthing switch can be closed only if the downstream breaker is open (b) Upstream breaker can be closed only if the downstream earthing switch is open.

(5)   Bus earthing switch shall be interlocked to close only if incomers, bus coupler and outgoing feeder breakers are isolated.

(6)   Breakers racking in/out shall be permitted only with the breaker door closed.

(7)   Switchgears shall be specified to be provided with barriers/enclosures with a degree of Ingress Protection of IP2X or better according to IEC-60364-4-41 for parts that are to be accessed with live circuits or with live parts behind the barriers.

(8)   Automatic changeover shall be inhibited if the incomer or bus coupler breakers have tripped due to electrical faults.

(9)   Caution labels shall be affixed for the following

  (a)   Cable compartment doors of all source breakers shall have "CAU-TION: ISOLATE INCOMING POWER AT FEEDER No. XX FROM SWITCHGEAR XX BEFORE OPENING THIS COVER".

  (b)   All feeders shall have "CAUTION: HIGH VOLTAGE" caution labels.

      (c)   Cubicles with dual incomers shall be provided with "CAUTION: DUAL FEED. ISOLATE INCOMING POWERS AT FEEDER No. XX FROM SWITCHGEAR XX AS WELL AS AT FEEDER No. XX FROM SWITCHGEAR XX BEFORE OPENING THIS COVER".

(10)   Breakers and starter disconnectors shall be provided with padlocking facility. No locking shall be possible while in service as this will defeat the emergency isolation provision.

(11)   Switchgears shall be specified to be provided with interlock to prevent inserting breaker/motor starter cubicles of ratings and schematics different from the original one. That is, each and every rating and schematic type should be unique and not interchangeable. This will address two important safety related issues.

      (a)   If exchanged with under rated breaker starter would result in starter/breaker failure. If under rated in terms of full load current, short circuit rating etc. the breaker/starter would not be able to carry the required load/short circuit current and this would result in component failure with explosion and fire possibility.

      (b)   If exchanged with rating higher than the actual starter/breaker the protection would fail to protect the connected load including the cable if the setting is higher than the connected load and cables ratings. In most cases this has been the cause of failures. Often starters and breakers are exchanged without even checking the breakers/starters' settings (in some cases the setting range itself will not be suitable for the connected load-often the minimum far exceeds the protected equipment rating).

(12)   Switchgears shall be provided with electrical insulating mats in front of the switchgears.

(13)   Switchgear rooms shall be displayed with first-aid charts and procedures for victims of electrical shock.

4.3   Motors

Motors do not cause too much unsafe conditions. However the following safety aspects shall be taken care.

(1)   Motor protections should be set based on actual driven equipment load and not based on motor full load current so that the driven equipment abnormalities can also be sensed and protected against damage. Motors over load and over current protections should be set giving only reasonable margin. Excessive margin will cause danger in situations of mechanical failure of driven equipment.

(2)   Guards shall be installed to protect against rotating parts.

(3)   Motors frames shall be earthed.

(4)   Motors shall be suitable for the classification of the area in which it is installed.

4.4   Lighting and Small Power Distributions

Following safe practices shall be followed:

(1)   Small power outlet circuits shall be provided with earth leakage circuit breakers set to operate at 30 mA to clear the fault within 0.04 s at 5 times the operating current.
(2)   Lighting circuits shall be protected by 300 mA earth leakage breakers to prevent fire possibilities due to leakage current.
(3)   Socket outlets shall be provided with interlocking switch such that plugging in/out is possible only when the switch is OFF.
(4)   The distribution PE conductor shall be adequately sized as recommended in IEC-60364-4-41 Section 543.

4.5   Cables

Cables play major role in industrial fire accidents often causing enormous loss to men, materials and installations. Following guidelines shall be followed

(1)   Cables shall be designed to have current rating with at least 10% safety margin over maximum demand after considering de-rating effects due to ambient temperature, grouping factors (number of cables in one cable tray), number of tiers of cable trays, method of installation (in air, buried, in conduit etc.) type and method of installation (three core, single core, trefoil or flat formation etc.), circulating current in the screen (for screens earthed at both ends). Guidelines given in IEC 60364 shall be followed.
(2)   Cables are protected by, in most cases, by providing overload protection for the connected load and as the cables are normally rated higher than the load current, the cables are automatically protected. However, if there is mismatch between the load and cable sizing the cable will not be protected. This is the main cause of most cable failures. The overload protection shall ensure that it will definitely operate at currents less than or equal to 1.45 times the continuous current carrying capacity of the cables according to IEC 60364-4-43. Since fuses are rated in standard sizes, their ratings, except in few cases, will lower far than the cable ratings. Also the fuses have to be selected not to fuse out during peak demand such as largest motor starting. Due to these reasons, there is possibility that the cables are not protected and hence care should be exercised in selecting the fuses.
(3)   Cables shall be protected with fire coatings at entry points into the switchgear and also at transition points.
(4)   Cables shall be installed well within the allowable bending radius having 25% margin recommended by the manufacturer.

4.6   *Earthing*

Earthing plays an important role in electrical systems. Earthing can be classified into three groups (a) System neutral Earthing (b) Equipment frame earthing (earthing of conductive non-current carrying part of electrical equipment (c) extraneous conductive parts as defined by IEC (Conductive parts not forming part of the electrical equipment) (d) Earth grid.

Following safe practices shall be followed in electrical earthing:

(1)   Earth grid design for major installations shall have a value less than 1 Ω (preferably less than 0.5 Ω).

(2)   Step and touch voltages shall be well within allowable tolerable limits given in IEEE 80.

(3)   Where fences are installed the earth grid shall extent beyond the fence and fence shall be earthed to the earth grid adequately at regular intervals (at least every 25 m).

(4)   Touch potentials of fences is usually an area of concern, as normal design can't ensure safe limits. Design should have adequate measures (add earth rods at the perimeter or follow unequal spacing design) to achieve safe limits.

(5)   Local Earth mats should be considered, at points in the switchyard (such as isolator and earth switch operating locations, local control panels etc.), where operator will be present for operating and a fault create hazardous voltage differences. The local earth mat and equipment including panels and operating handles shall all be bonded together electrically and then connected to the earth grid to eliminate major voltage differences.

(6)   System neutral shall be earthed using mechanically rugged conductors or else they should be protected using pipes and conduits having adequate strength to protect against possible mechanical damages.

(7)   Surge arresters shall be connected to earth using shortest path directly from the surge arrester without making any loop increasing the inductance.

(8)   All major equipment (MV and above) frames shall be provided with earthing at two points which are at diametrically opposite points.

(9)   Earthing conductor shall be adequate enough to withstand the rated fault current without melting and losing earthing integrity. A safety margin of at least 25% shall be followed for the temperature rise with respect to the melting temperature.

(10)   All exposed non-current carrying parts of equipment and enclosures and major structures shall be earthed.

(11)   The minimum size of the protective conductor shall be as given in IEC 60364-4-41 Section 543.

4.7   Lighting and Small Power Distribution Boards

(1)   Residual current circuit breakers rated at 30 mA shall be used in circuits supplying power to all socket outlets that will be used for temporary portable equipment.

(2)   Lighting circuits that are vulnerable for fire accident due to earth leakage currents, particularly in hazardous areas, shall be provided with 300 mA RCCBs.

(3)   For the purpose of isolation of circuits, in particular in classified hazardous areas, DBs shall be specified to be provided with circuit isolation facility with locking (pad locking or special locking provision).

(4)   Lighting fixtures and other equipment shall be affixed with a caution label "CAUTION: WAIT FOR XX MINUTES AFTER TURNING OFF THE POWER SUPPLY BEFORE OPENING THE COVER" for all those devices and equipment containing energy retaining properties/possibilities to allow for the discharge/decrease of the stored energy to a safe level well within the time indicated.

4.8   AIS

AIS layout shall address clearances relating to

(a)   Maintenance practices with respect to equipment replacements.

(b)   Safety to persons shall normally be achieved by provision of adequate Safety Clearances (SC) to live parts taking into account the need for construction, modification, maintenance and vehicular and pedestrians access. SC shall be clearly marked showing the dimensions in the sectional view drawings of AIS both along and perpendicular to the bay axis. The Drawing shall indicate both the section clearances according to the recommended values and that what is actually provided. The minimum clearances shall be as indicated in the Tables 3 [2] and 4 [3].

4.9   Batteries

Following safety related issues shall be addressed in the batteries General Safety Hazards:

(1)   Energy Thresholds. Energy exposure levels shall not exceed those identified in the following list unless appropriate controls are implemented:

   –   AC: 50 V and 5 mA
   –   DC: 100 V

Table 3 Minimum clearance for AIS (Voltage Range I) as per IEC61936-1

Highest voltage for installation	Rated short-duration power frequency withstand voltage	Rated lightning impulse withstand voltage[a]	Minimum phase-to-earth and phase-to-phase clearance	
Um r.m.s	Ud r.m.s	Up 1.2/50 μs (peak value)	Indoor installations	Outdoor installations
kV	kV	kV	mm	mm
3.6	10	20	60	120
		40	60	120
7.2	20	40	60	120
		60	90	120
12	28	60	90	150
		75	120	150
		95	160	160
17.5	38	75	120	160
		95	160	160
24	50	95	160	
		125	220	
		145	270	
36	70	145	270	
		170	320	
52	95	250	480	
72,5	140	325	630	
123	185[b]	450[b]	900	
	230	550	1100	
145	185[b]	450[b]	900	
	230	550	1100	
	275	650	1300	
170	230[b]	550[b]	1100	
	275	650	1300	
	325	750	1500	
245	275[b]	650[b]	1300	
	325[b]	750[b]	1500	
	360	850	1700	
	395	950	1900	
	460	1050	2100	

[a]The rated lightning impulse is applicable to phase-to-phase and phase-to-earth
[b]If values are considered insufficient to prove that the required phase-to-phase withstand voltages are met, additional phase-to-phase withstand tests are needed

Table 4 Minimum clearance for AIS as per IEEE 1427

Maximum system (c) voltage phase-to-phase	Basic BIL (c)	Minimum phase-to-ground (d, f) clearances	Minimum phase-to-phase (d, e, f) clearances
(kV, rms)	(kV, crest)	mm	mm
1.2	30	57	63
	45	86	95
5	60	115	125
	75	145	155
15	95	180	200
	110	210	230
26.2	150	285	315
36.2	200	380	420
48.3	250	475	525
72.5	250	475	525
	350	665	730
121	350	665	730
	450	855	940
	550	1045	1150
145	350	665	730
	450	855	940
	550	1045	1150
	650	1235	1360
169	550	1045	1150
	650	1235	1360
	750	1325	1570
242	650	1235	1360
	750	1425	1570
	825	1570	1725
	900	1710	1880
	975	1855	2040
	1050	2000	2200
362	900	1710	1880
	975	1855	2040
	1050	2000	2200
	1175	2235	2455
	1300	2470	2720
550	1300	2470	2720
	1425	2710	2980
	1550	2950	3240
	1675	3185	3500
	1800	3420	3765

(continued)

Table 4 (continued)

Maximum system (c) voltage phase-to-phase	Basic BIL (c)	Minimum phase-to-ground (d, f) clearances	Minimum phase-to-phase (d, e, f) clearances
(kV, rms)	(kV, crest)	mm	mm
800	1800	3420	3765
	1925	3660	4025
	2050	3900	4285
	2300	4375	4815

(a) Clearances shown are based on a 605 kV/m flashover gradient
(b) Switching surge conditions normally govern for system voltages above 242 kV
(c) Values for maximum system voltages and BIL levels are from Tables 1 and 2 of IEEE Std 1313.1-1996, except for the 1.2 and 5 kV system voltage and the 30, 45, 60, 75, and 2300 kV BIL values
(d) For specific equipment clearance values, see relevant apparatus standards
(e) Phase-to-phase clearances shown in this table are metal-to-metal clearances not bus-to-bus centerlines

(2) Battery Risk Assessment. Prior to any work on a battery system, a risk assessment shall be performed to identify the chemical, electrical shock, and arc flash hazards and assess the risks associated with the type of tasks to be performed.
(3) Battery Room or Enclosure Requirements

  – Personnel Access to Energized Batteries. Each battery room or battery enclosure shall be accessible only to authorized personnel.
  – Illumination. Employees shall not enter spaces containing batteries unless illumination is provided that enables the employees to perform the work safely.

(4) Abnormal Battery Conditions. Instrumentation that provides alarms for early warning of abnormal conditions of battery operation, if present, shall be tested annually.

4.10 Exposed Live Parts

Exposed live parts in electrical systems and equipment shall be protected against direct contact by one of the following methods according to IEC 60364-4-41.
(1) The barriers/enclosures with a degree of protection of at least IP2X.
(2) Live parts shall be completely covered with insulation which can only be removed by destruction.
(3) By providing obstacles.
(4) By placing it out of reach distance.
(5) Protection by residual current devices.

   If inspection of exposed live part of electrical equipment is required when it is live, at least two layers of bolted barriers shall be provided to prevent direct contact. It shall be possible to remove these barriers only with special tools.

Monitors (leakage and surge counter) for surge arresters should be installed such that the cable between surge arrester and the monitor is well insulated to avoid touch potential and it is preferable to locate them at out of reach distance but at the same time the meter can be read standing on the ground. The cable between surge arrester and the monitor will reach several kV during a lightning flash because of very high rate of rise of the discharge current flowing through the inductance of the meter. It is recommended that cable of adequate voltage rating is procured along with monitor from the same vendor.

5   Typical ELSOR REPORT

5.1   Preliminary and Detailed Design Reviews

ELSOR review shall be done in two stages for Preliminary Design and Detailed Designs. As ELSOR review covers wide range of issues the applicable points for the ELSOR review in each stage will also be different, as the designs don't get finalized in just one review. Following guidelines may be used in general to select the points of review of Preliminary and Detailed Designs.

5.1.1   Preliminary Design ELSOR Review

In Preliminary Design the system design will be finalized and fault levels also selected. Hence the following points relating to the design shall be reviewed in the Preliminary Design Stage

(1)   Equipment sizing calculation.
(2)   Fault Study, Load Flow Study, Motor Starting Study and Harmonic Analysis.
(3)   Equipment layouts.
(4)   Reliability Design based on redundancy.
(5)   Design Basis.
(6)   Specifications.

5.1.2   Detailed Design ELSOR Review

Detailed design review shall include all other points that are not covered in Preliminary Design Review, including but not limited to,

(1)   Earthing and Lightning Protection Designs.
(2)   Single line diagram relay protections.
(3)   Relay settings.
(4)   Datasheets.
(5)   Equipment Sizing (In the cases where changes in ratings have been done).
(6)   Calculations
(7)   Cable routing etc.

5.2 ELSOR Report Contents

The contents of a typical report shall have the following sections to make the report systematic and easy to comprehend.

(1) Project Introduction.
(2) List of Abbreviations and Definitions.
(3) ELSOR Operability Review.

    (a) ELSOR Objectives.
    (b) Scope of Review.
    (c) Methodology.

(4) ELSOR Findings.

    (a) ESSID Findings.
    (b) ESTOS Findings.

(5) Conclusions.
(6) References.
(7) Appendices.

    (a) List of Participants along with their role in the review.
    (b) Diagrams and ESSIS and ESTOS Worksheets.

5.3 Typical ELSOR Worksheet

5.3.1 Typical ESSID Worksheet

A typical ESSID worksheet with an example is presented below, Table 5. The work sheet shall be prepared for each part of the system that is identified and listed for review. Within a part, each equipment shall be reviewed using Guidewords listed in this guideline as minimum. Additional points may be added by the review team as deemed fit and relevant for the document that is reviewed.

5.3.2 Typical ESTOS Worksheet

A typical ESTOS worksheet with an example is presented below. The work sheet shall be prepared for each part of the system that is identified and listed for review. Within a part each equipment shall be reviewed using Guidewords listed in this guideline as minimum. Additional points may be added by the review team as deemed fit and relevant for the document that is reviewed (Table 6).

    Detail ESSID and ESTOS reports of one oil field electrical engineering can refer attachments on this book.

Table 5 Example of ESSID worksheet

SL	Doc	Equipment	Design aspect	Guideword	Lacuna	Recommendation	Action party	Remarks/Notes
1	11 kV incomer	Feed transformer	51 pickup setting	Setting adequacy	Set too low at 105% without adequate margin for short time loading conditions	Recommended to set 1t 125%		Setting to be increased and co-ordination to be rechecked
2	Switchgear layout	Panels	Rear access	Adequacy for maintenance activities	900 mm rear access is adequate	Nil		Switchgear is front access type and hence rear access is not required

Table 6 Example of ESTOS worksheet

SL	Doc	Equipment	Design aspect	Guideword	Lacuna	Recommendations	Action by	Notes/Remarks
1	Transformer specification	Transformer	Fire/Explosion-Others	Protection against pressure rise	OLTC not provided with pressure relief valve	Relief valve of adequate capacity shall be provided		
2	Switchgear specification	Switchgear breaker/modules	Ratings	Interchangeability safety	Not specified	Interchangeability shall be specified only between same ratings and type of breakers and starter modules		

References

1. IEC 61439-1-2011, Low-voltage switchgear and controlgear assemblies – Part 1: General rules, Section 11.10.
2. IEC 61936-1-2014, Power installations exceeding 1 kV a.c. – Part 1: Common rules, Section 5.4.
3. IEEE 1427-2006, IEEE Guide for Recommended Electrical Clearances and Insulation Levels in Air-Insulated Electrical Power Substations, Section 6.3.

Chapter 2
Safety Analysis of Power Distribution Engineering in International Oil Field

International oil and gas fields are generally located in remote areas without national utilities system support, the design safety of power supply and distribution in international oil field surface engineering is particularly important. Only when the design safety is achieved then the safety of production and operation can be guaranteed. There are many mandatory technical requirements in industry and national standards for oil field power system engineering in domestic, but seldom and dispersed in different international standards. This chapter summarizes the main technical safety requirements for oil field power system engineering safety, wish to provide references for similar engineering in both domestic and international project.

In international oil and gas field ground engineering design, the concept relating to design safety mainly include load classification and load calculation, short-circuit current calculation, system configuration, overvoltage protection, electrical safety and fire prevention distance, electrical safety interlocks and phase color identification, etc. Load classification and load calculation directly relates to the owner's approval of the system and whether the equipment capacity meets the requirements of safety and reliability. The calculation results of short-circuit current directly affect the selection of equipment tolerance value and relay protection setting, and have comprehensive relation to the safe operation of the system and equipment. Due to the different grounding model of medium voltage distribution systems in domestic and abroad, the typical lightning arrester parameters in domestic may not be able to meet the requirements of international projects. Therefore, it is necessary to understand the selection of lightning arrester parameters in detail to ensure the safe operation of equipment. There are still some differences between international and domestic standards on the safety distance of outdoor distribution devices, especially the transformer fire spacing. These need to be fully understood and mastered in international projects to avoid the failure to meet the safety distance requirements within the codes based on domestic standards. In addition, phase color identification in domestic and foreign standards is compared, which it has the direct impact on the safety operations for design reference.

© Petroleum Industry Press 2022
K. Ma et al., *International Oilfield Surface Facilities: Safety Analysis for Electrical Design*, https://doi.org/10.1007/978-981-16-3104-7_2

In the design process of power system for oil field surface engineering, they are systematically stipulated in equipment selection and system calculation in domestic standards, which are detailed enough and not be mentioned in this book. This chapter only highlights the relevant key points of technical disputes frequently appeared in international projects for the reference of peers. With regarding to the technical problems related to safety which are often encountered in the design of power supply and distribution system, and for which there are no relevant provisions in international standards but there are provisions in national standards, are also listed here for design reference. It expected that this chapter would be served as a quick index of electrical design safety principles for international s oil field surface engineering.

1   Load Classification

1.1   Load Classification for Power System

As per China national standard GB 50052 [1] mandatory technical requirements, load classification shall be as per the importance degree, scale, load capacity and loss or influence in personal safety and economy caused by the interruption of power supply. The below technical requirements shall be followed:

(1)   If one of the following conditions is complied, it shall be Class I level load.
(a)   Interruption of power supply will result in personal injury.
(b)   Interruption of power supply will result in great economic losses.
(2)   In load of Class I level, when the interruption of power supply will cause casualties or major equipment damage or poisoning, explosion and fire, as well as in the particularly important place is not allowed to interrupt power supplying, shall be considered as the load of special importance in the Class I level load.
(3)   If the interruption of power supply will cause great loss (less than Class I) in economy, it shall be as Class II level load.
(4)   Those are not belong to the class I and II level load shall be the Class III load.

1.2   Load Classification for Oil Field

As per China national standard GB 50350 [2], oil field power loads can be classified as Class I, Class II and Class III, three levels and the typical load classification are as followings:

(1)   Class I level load is as follows:

Oil field: with capacity equal to or higher than 30×10^4 t/a of central processing facilities, field oil depot (pipeline transportation), light hydrocarbon depot.

Pipeline: crude oil export station, terminal station, pressure reducing station, intermediate heat pump station with stable pressure, compressor station with power as the driven source or other driven source with higher demand of power consumption.

The automatic control system and communication system of the cut-off valve station of oil and gas transmission line, crude oil transfer station and gas transfer station. The emergency shunt down valve and emergency lighting of oil and gas transfer station should be the super important load in class I level load.

(2)   Class II level load is as follows:

Oil field: with capacity less than 30×10^4 t/a central processing facilities, field oil depot (railway transportation), crude oil stability station, transfer station, water injection pump station with 10(6) kV motor, sewage treatment station, crude oil dehydration station, compressor gathering station, gas injection station, mechanical driven oil wells, etc.

Pipeline: Intermediate heat pump station with pressure transferring, heating station.

(3)   Class III level load is as follows:

Oil Wells with natural flow and gas Wells, remote and isolated mechanical production Wells, remote gas transfer stations, manifold station, water distribution stations and Independent cathodic protection station should be Class III.

As per domestic national standard, load classification is the basis for power system design, and oil field load classification is the fundamental of further power supply and distribution.

There is no specific load classification of international projects equal to domestic oil field load level. Generally, all the oil field stations adopt double circuit power supply at least in international oil field with exception of those unimportant production wells outside the station with the loop power supplying. Comprehensive assessment of power supply stability and its influence of power failure should be included in field development plan before the power supply configuration selected.

1.3 Load Classification as Per IEC

In the international standard, the load classification is only divided into continuous load, intermittent load, standby load and emergency load. For different production units, there are some examples of double power supply, double circuit and special circuit power supply, which are not completely in consistent with the domestic mandatory engineering regulations.

IEC 60364-1 [3], provides one load classification for the continuity of power supply according to change-over time:

- No-Break: an automatic supply which can ensure a continuous supply within specified conditions during the period of transition, for example as regards variations in voltage and frequency;
- Very Short Break: an automatic supply available within 0.15 s;
- Short Break: an automatic supply available within 0.5 s;
- Medium Break: an automatic supply available within 15 s;
- Long Break: an automatic supply available in more than 15 s.

Based on practical engineering experiences, international standards pay more attention to the unification of economy and reliability in system design, and have different requirements for different engineering projects. At present, there are no compulsive stipulations for load classification abroad. The IEC 60364-1 only makes requests for classification of automatic switchover based on the demand of switching time in low voltage loads and divides the automatic switchover into five class.

When selecting automatic transfer devices in international projects, the requirements of change-over time can be referred to these five categories of IEC 60364-1. The change-over time of low-voltage system ATS (Auto Transfer Switching) is in accordance with IEC standard, and the default standard time should be 0.15, 0.5 s.

DL 5136 [4] defines high speed change-over device shall be not higher than 100 ms (0.1 s), which is not completely in accordance with IEC and is mainly used in important places of thermal power plants in China. The transfer time of 100 ms can meet the technical requirements of relevant equipment in general.

2 Load Calculation

2.1 Load Calculation in Oil Field

Industrial and Civil Power Distribution Design Manual in China specified several typical calculation methods in domestic, including the demand coefficient method, utilization coefficient method, unit power method, etc., international oil field project load calculation normally take into account the equipment's load coefficient, load continuity and is a comprehensive calculation method [5].

Typically, international oil field design projects calculate and list power loads according to the three types of continuous load, intermediate load and standby load in order to determine maximum normal operating load, peak load and minimum capacity of emergency generators.

Typical calculations are as followings:

$$LF = P_{shf}/P \tag{1}$$

$$P = P_{shf}/\eta \tag{2}$$

$$Q = P \times \tan\theta \tag{3}$$

$$S = \sqrt{P^2 + Q^2} \tag{4}$$

$$P_{Max.OP} = \sum_c P_c + (0.3 \sum P_I) \text{ or } P_{LSI} \text{ (Apply the larger one)} \tag{5}$$

$$Q_{Max.OP} = \sum Q_c + (0.3 \sum Q_I) \text{ or } Q_{LSI} \text{ (Apply the larger one)} \tag{6}$$

$$S_{Max} = \sqrt{P_{Max.OP}^2 + Q_{Max.OP}^2} \tag{7}$$

$$P_{Peak} = \sum_c P_c + (0.3 \sum P_I) \text{ or } P_{LSI} \text{ (Apply the larger one)}$$
$$+ (0.1 \sum P_s) \text{ or } P_{LSS} \text{ (Apply the larger one)} \tag{8}$$

$$Q_{Peak} = \sum_c Q_c + (0.3 \sum Q_I) \text{ or } Q_{LSI} \text{ (Apply the larger one)}$$
$$+ (0.1 \sum Q_s) \text{ or } Q_{LSS} \text{ (Apply the larger one)} \tag{9}$$

$$S_{Peak} = \sqrt{P_{Peak}^2 + Q_{Peak}^2} \tag{10}$$

LF—Load Factor;
P_{shf}—Shaft Power, kW;
P—Active Power, kW;
Q—Reactive Power, kVar;
S—Apparent Power, kVA;
P_c—Continuous Active Load, kW;
P_I—Intermittent Active Load, kW;
P_s—Spared Active Load, kW;
P_{LSI}—The largest Single Intermittent Equipment Active Rating, kW;
P_{LSS}—The largest Single Spare Equipment Active Rating, kW;
Q_c—Continuous Reactive Load, kVar;
Q_I—Intermittent Reactive Load, kVar;
Q_s—Spared Reactive Load, kVar;
Q_{LSI}—The largest Single Intermittent Equipment Reactive Rating, kVar;
Q_{LSS}—The largest Single Spare Equipment Reactive Rating, kVar;
$P_{Max.OP}$—Maximum Active Operation Rating, kW;
$Q_{Max.OP}$—Maximum Reactive Operation Rating, kW;
$S_{Max.}$—Maximum Apparent Rating kVA;
P_{Peak}—Maximum Active Operation Rating, kW;
Q_{Peak}—Maximum Reactive Operation Rating, kW;
$S_{Peak.}$—Maximum Apparent Rating kVA;

η—Efficiency;
Cosθ—Power Factor (PF).

2.2 De-rating Factor

The load list shall be compiled according to the load data provided by the process, mechanical, instrument and other related majors disciplines. While in the processing of designing the oil field facilities, the process engineer has already taken the load fluctuation factor into account. For example, actual oil field equipment processing capacity of a 1.0 million ton will consider the production fluctuation to 1.2 million tons to accommodate inrush operation. Therefore, when selecting load factor and simultaneous factor, process requirement shall be considered in the total load calculation of the overall oil and gas field. According to oil and gas field production and operation experience, typical simultaneous factor of main substation is 0.9, and substation of terminal substation is 0.9–0.93, the simultaneous factor of each unit area of the station is 0.93–0.95, and the simultaneous factor of single well station is 0.6–0.7 according to practical operation condition and the amount of the wells. In brief, the selection of simultaneous factors is considered step by step and should be objectively selected according to the characteristics of load and the experience of oil and gas field production and operation.

During the actual calculation process, the maximum utilization rate of motor should be considered. In the absence of effective information of the motor, the load factor, efficiency and power factor of the motor are set according to the following Tables 1 and 2.

For non-motor load, typical data can refer to Table 3.

Load Factors of continuous, intermittent and spare loads can refer to "Handbook of Electrical Engineering for Practitioners in the Oil, Gas, & Petrochemical Industry" [6] and details can refer to Table 4.

3   Short-Circuit Calculation

Short circuit current of power system is an important section in system design and equipment selection. ETAP or EDSA software is often as the tools for calculation in international engineering. In engineering application, if the short-circuit current calculation is too conservative, it may cause unnecessary higher investment, on the other hand, the system may have potential risk if the value specified small. This chapter does not describe the specific theory of short-circuit current calculation, but only briefly describes the differences in calculation standards and different calculation concepts at home and abroad.

Table 1 Typical motor data 1

Rated power (kW)	Efficiency	Power factor
0.18	0.60	0.63
0.25	0.66	0.74
0.37	0.68	0.74
0.55	0.73	0.76
0.75	0.75	0.76
1.10	0.78	0.78
1.50	0.79	0.79
2.20	0.81	0.82
3.00	0.83	0.81
4.00	0.85	0.82
5.50	0.86	0.84
7.50	0.87	0.85
11.00	0.88	0.84
15.00	0.89	0.85
18.50	0.91	0.86
22	0.92	0.86
30	0.92	0.87
37	0.92	0.87
45	0.92	0.88
55	0.93	0.88
75	0.93	0.88
90	0.94	0.89
110	0.94	0.89
132	0.94	0.89
160	0.95	0.89
185	0.94	0.89
200	0.95	0.89
225	0.94	0.89
250	0.96	0.90
280	0.95	0.90
315	0.96	0.90
355	0.95	0.86
400	0.95	0.86
450	0.95	0.86
500	0.95	0.87
630	0.96	0.87

(continued)

Table 1 (continued)

Rated power (kW)	Efficiency	Power factor
710	0.96	0.87
800	0.96	0.87
900	0.96	0.87
1000	0.96	0.87
1120	0.96	0.87
1250	0.96	0.88
1400	0.96	0.88
1600	0.97	0.88
1800	0.97	0.88

Note 315 kW and above are normally medium voltage motor, and rated voltages are normally 6 and 10 kV

Table 2 Typical motor data 2

Rated power (kW)	Load factor	Efficiency	Power factor
$x < 15$	0.70	0.85	0.73
$15 \leq x < 45$	0.75	0.91	0.78
$45 \leq x \leq 150$	0.83	0.93	0.82
>150	0.85	0.94	0.91

Table 3 Typical non-motor equipment data

Load type	Efficiency	Power factor
Packaged equipment	0.91	0.95
Electrical heater	1.00	0.95
IGBT controlled industrial heater	0.93	0.82
AC and DC UPS	0.85	0.85
Lighting	0.90	0.90

Table 4 Load factors of continuous, intermittent and standby loads

Type of project	D_c for C_{sum}	D_i for I_{sum}	D_s for S_{sum}
Conceptual design of a new plant	1.0–1.1	0.5–0.6	0.0–0.1
Front-end design of a new plant (FEED)	1.0–1.1	0.5–0.6	0.0–0.1
Detail design in the first half of the design period	1.0–1.1	0.5–0.6	0.0–0.1
Detail design in the second half of the design period	0.9–1.0	0.3–0.5	0.0–0.2
Extensions to existing plants	0.9–1.0	0.3–0.5	0.0–0.2

3.1 Calculation Standards

There are mainly four reference standards for short-circuit calculation including IEC, ANSI/IEEE, Russian standard (GOST), Chinese standard (GB). In ETAP short circuit calculation can apply IEC, ANSI and GOST standards. Details for calculation standards can refer to the followings:

ANSI/IEEE Std C37.5™, IEEE Guide for Calculation of Fault Currents for Application of AC High-Voltage Circuit Breakers Rated on a Total Current Basis.

IEC 60909, Short-circuit currents in three-phase a.c. systems.

IEC 61363-1: Electrical installations of ships and mobile and fixed offshore units—Part 1: Procedures for calculating short-circuit currents in three-phase a.c.

IEEE Std 141™, IEEE Recommended Practice for Electric Power Distribution for Industrial Plants (IEEE Red Book™).

IEEE Std 241™, IEEE Recommended Practice for Electric Power Systems in Commercial Buildings (IEEE Gray Book™).

IEEE Std 242™, IEEE Recommended Practice for Protection and Coordination of Industrial and.

Commercial Power Systems (IEEE Buff Book™).

IEEE Std 551™, IEEE Recommended Practice for Calculating AC Short-Circuit Currents in Industrial and Commercial Power Systems (IEEE Violet Book™).

IEEE 3002.3 Recommended Practice for Conducting Short-Circuit Studies and Analysis of Industrial and Commercial Power Systems.

By comparing the ANSI and IEC standards for short-circuit calculations, there are two major difference with each other from modeling of equipment to calculation methods. One common question from electrical engineers is, which one tends to provide more conservative results? There is no general answer to this question, since it depends on the electrical system under consideration. It should be pointed out that as both calculation methods are based on approximate models, choosing the method that provides the largest short-circuit current in the system is not very relevant. This is because one of the main purposes of short circuit calculations is, in fact, to size, or verify, ratings of equipment. The selected short-circuit current calculation method must be in compliance with the standard upon which the equipment has been manufactured [7].

Although significant effort has been made to harmonize rating frame for higher voltage circuit breakers in the new standard, the rating frame and testing requirements for bus, circuit breakers, fuse, switch, etc., are not quite in agreement between ANSI and IEC standards. These standards are created to go hand in hand with the corresponding standards for equipment ratings. Therefore, if a system contains equipment in compliance with ANSI standards, then the ANSI standard short-circuit calculation method must be selected to evaluate this equipment. This is also true for the equipment in compliance with IEC standards.

In ANSI short-circuit calculation, equipment impedance values are mainly based on parameters provided by manufacturers, with certain tolerance applied to achieve conservative fault current values. In IEC short circuit calculations, a correction factor is applied to synchronous machines and transformers to account for normal operating conditions.

Transformer modeling—In ANSI standard, the transformer impedance values provided by the manufacturer are used in the calculation. To take into account the possible inaccuracy of these parameters, when they are not obtained from field testing of actual equipment, additional impedance tolerance may apply. In IEC standard, an impedance correction factor (K_T) is applied to transformer impedance to take into account prefault operating conditions, including transformer taps. The correction factor K_T is calculated differently based on the transformer being a network transformer or a power station unit transformer.

Machine modeling—According to ANSI and IEC standards, all machines are modeled by a constant voltage source behind an impedance. The two methods differ in how the machine impedance is utilized. In ANSI short-circuit calculation, induction machine impedance is calculated based on motor locked-rotor impedance multiplied by a factor, defined as the ANSI multiplication factor. The ANSI multiplication factor is applied to take into account machine operating conditions and effects of motor feeder cable and overload heater, and its value varies depending on machine size and speed. The synchronous machine impedance is based on parameters provided by manufacturers. It should be noted that when performing device duty calculation for generator circuit breakers per IEEE Std C37.013™-1997, a detailed synchronous generator model must be used, which includes machine sub-transient and transient impedance and time constant to accurately account for machine AC and DC decays.

In IEC short-circuit calculations, the impedance of synchronous generator and compensator is adjusted by a factor (KG) to account for prefault operating condition and excitation of the machine. If a generator is a part of a power station unit, a different adjustment factor is used. For induction motors, the locked-rotor impedance is used in the calculation without any adjustment.

3.2 Calculation Principles

The system operation mode in short-circuit calculation shall be as per the mode with the largest short-circuit current. Transient parallel of incomers shouldn't be considered [8].

When determining the short circuit current, it shall follow the normal operation mode with the maximum possible short circuit current, and shall not follow the wiring mode with all possible parallel operation in the switching process.

Shell DEP 33.64.10.10 ELECTRICAL ENGINEERING DESIGN [9] requirements are that short circuit rating of equipment and cables, including short circuit manufacturing and the associated circuit switching equipment's breaking capacity shall be based on the connection mode of parallel operation.

DEP 33.64.10.10 also stipulated that the short-circuit capacity selection of medium-voltage equipment should consider 10% margin on the basis of the calculation results. Therefore, in the actual project, the short-circuit calculation cannot be simple in accordance with the international standard, and the technical provisions of bidding documents must be strictly followed. In the short circuit calculation stage.

In the actual project operation, the short-circuit current calculation is more feasible according to DEP 33.64.10.10, but if the type, quantity and power of the load are clear, the simulation calculation can be carried out according to the actual working conditions.

4 System Configuration

Power system configuration is the most important part in power system design. The system Configuration is also related to reliability, flexibility and safety of the whole power system and has a heavy influence on the selection of electrical equipment, distribution arrangement, relay protection and control mode. It is necessary to comprehensively analyze the influencing factors, and reasonably determine the main system architecture through the comparison of technology and economy.

There are mandatory power system configuration provisions based on different load classes.

4.1 Mandatory Technical Requirement for System Configuration in GB 50052 [1]

Double power supplies shall be provided for Class I load. The two power supplies should from different substations or power plants. When two power sources are respectively drawn from different bus sections of the same substation, the substation shall have at least two power supplies and at least two main transformers in parallel operation.

For the especially important load in Class I level load, in addition to two power supplies, emergency power supply should be provided, and other loads should not be connected to the emergency power supply system.

Two circuits shall be provided for Class II level loads. If it is impossible to provide two circuits, a dedicated overhead line or cable can be used for power supply only when process system has power-off safe design or emergency power supply available.

4.2   Technical Requirement for System Configuration in GB 50049 [10]

Power system Configuration design, with except as per load classifications, can be also based on power loads capacity. GB 50049 specified two kinds of power system Configurations based on load ratings:

When the generator capacity of each section is 12 MW or less, it is advisable to use single bus or single bus.

When the generator capacity of each section is more than 12 MW, it is advisable to use double-bus or double-bus with bus tier.

4.3   Technical Requirement for System Configuration in API 540 [11]

With regarding to power system configuration design, API 540 specifies several typical types.

The size and importance of the power station will determine the type of bus arrangement utilized for the main electrical connections. Small stations (less than 10 MW) frequently have only a single main bus as shown in Fig. 1. Bus failures are not common, and fair reliability is obtained. It is necessary to shut down, however, when performing preventive maintenance to the main bus or when additional bus tie in are made to the main bus. Circuit breakers must be taken out of service to be worked on, but this problem can be minimized by using drawout-type breakers.

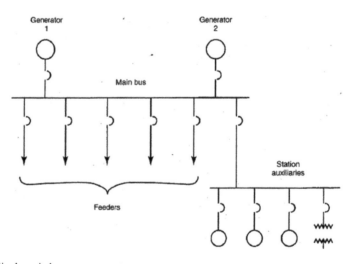

Fig. 1 Single main bus arrangement

These disadvantages can result in disruption to essential station auxiliaries, such as draft fans, preheaters, boiler feed pumps, air compressors, and lighting.

Unit construction, illustrated in Fig. 2, can be used in isolated power stations. With this construction, each genset is with its own main bus, its heater or its own boiler together with auxiliary loads are with its auxiliary bus. Normally, the tie circuit breaker between the main buses is closed but will open in the event of a fault on either bus. In effect, the arrangement operates much the same as two independent power stations tied together. The main buses are tied together during normal operation, so each side must be rated for the total fault duty resulting from both generators.

A synchronizing bus scheme, shown in simplified form in Fig. 3, is often used for a power station bus. This scheme offers a high flexibility to add or remove generators and loads. The reactors serve to limit the amount of current imposed on any one bus and to isolate voltage dips, to a degree, during faults. In this arrangement the loads on each bus are matched to the generating capacity on that bus to minimize the amount of load transfer through the synchronizing bus under normal operation.

The design of the power station bus arrangement should allow for future expansion, such as expanding a single or dual bus arrangement to a synchronizing bus arrangement. The design should also minimize the loss of generating capacity which would occur during a single fault or operating error.

Additionally, with reference to the oil well facilities in oil and gas field, especially in Middle East and North Africa oil field, the radial configuration is adopted commonly as Figs. 4 and 5, it is seldom seen in domestic oil field.

Domestic oil field power system structure is based on load classification from mandatory standards, that of in international oil field is based on the reliability and economy of the system, and there are corresponding guidelines to quantify the reliability calculation of the system.

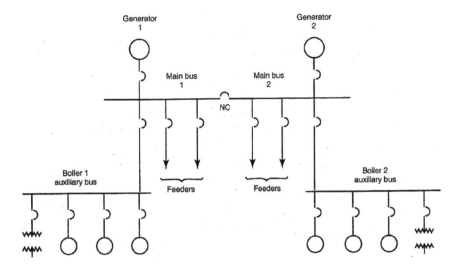

Fig. 2 Unit construction arrangement bus

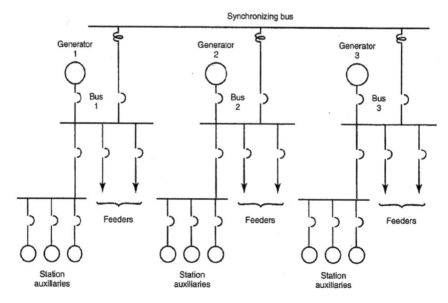

Fig. 3 Synchronizing bus arrangement

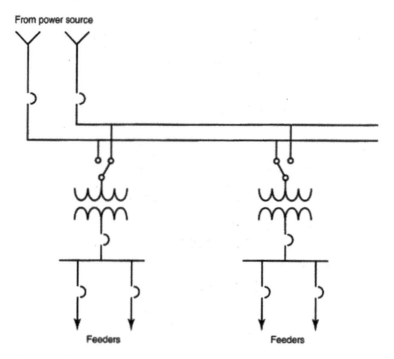

Fig. 4 Simple radial system

From power source

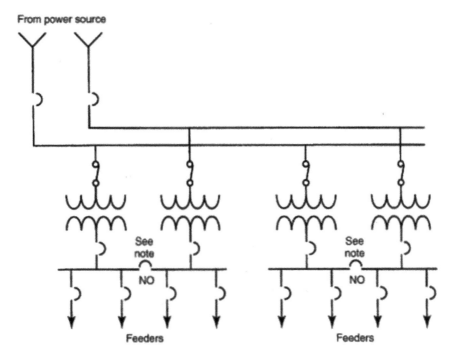

Fig. 5 Primary selective radial system

International standards generally do not have mandatory provisions for system design, which focuses on the unification of economy and reliability. However, for some overseas oil and gas field projects, the concept of 'Availability' is often adopted in electrical system design, and calculations are required during actual project execution. Relevant calculations can refer to IEEE 3006 and IEEE 3006.5. IEEE 3006 [12] gives reliability statistics of various types of equipment, while IEEE 3006.5 [13] gives calculation methods, procedures and steps. IEEE 3006 and IEEE 3006.5 containing historical collection of equipment accident rate which can be as for calculation basis, relying on scientific accumulative calculation results to determine the applicability and reliability of the system, has high recognition in the world. On the other hand, relative domestic mandatory rules lack quantitative analysis; it is not easily accepted during actual detail engineering stage. However, the calculation in accordance to applicability and reliability in relative standards abroad has universal persuasion.

5 Overvoltage Protection

Most of the oil field power system adopts medium voltage system, it is normally unearthed in domestic while most of the international oil field projects are resistor earthed and solid earthed. Different neutral earthing system lead to different withstand voltages of electrical equipment. In this chapter, the system overvoltage protection is mainly discussed through common equipment in electrical design of international oil fields, such as surge arrester parameter selection, SPD installation requirements and insulation coordination of medium voltage switchgear.

5.1 Surge Arrester Parameter Selection

5.1.1 Surge Arrester Parameter Selection as GB 50064 [14]

The neutral solid grounding is seldom adopted in China medium voltage system, while it is often used in international oil field projects. There are typical medium voltage system surge arrester parameters coordination in domestic power system engineering with the default un-effective earthed, so when select China typical surge arrester to abroad project, safety factor shall be checked since switchgear withstand voltages are different between effective and un-effective earthed system.

The MOA parameter selection in GB 50064 is shown in Table 5.

Oil field power system is with generators, rotating motors and other similar equipment, and the rated voltage of MOA, which is used for these rotation equipment protections, should be selected as followings:

- Ground fault clearance time is no more than 10 s, MOA rated voltage shall be equal or more than 1.05 times of rotation machine rated voltage.
- Ground fault clearance time is more than 10 s, MOA rated voltage shall be equal or more than 1.3 times of rotation machine rated voltage.
- The continuous operating voltage of the MOA shouldn't less than 80% of rated voltage.
- The rated MOA voltage for the neutral point of the rotation machine rated voltage shall not be less than $1/\sqrt{3}$ of the phase-earth MOA rated voltage.

5.1.2 Surge Arrester Parameter Selection as NB/T 35067 [15]

Technical specifications for MOA selection in NB/T 35067:

(1) MOA Rated Voltage (Ur):

$U_r \geq KU_T$.

K—Fault cleared in 10 s, K = 1.0; Fault cleared more than 10 s, K = 1.25;
U_T—Temporary Overvoltage.

With regards to MOA rated voltage selection, only consider the temporary overvoltage by earthing fault, load shedding and long line capacitance effect. Temporary overvoltage can refer to Table 6.

(2) Rated voltage of surge arresters for generator protection shall be as followings:

Table 5 MOA continuous operating voltage and rated voltage

Neutral earthing type		Continuous operating voltage (U_c kV)		Rated voltage (U_r kV)	
		Line to earth	Neutral	Line to earth	Neutral
Effective earthing	110 kV	$U_m/\sqrt{3}$	$0.27\ U_m/0.46\ U_m$	$0.75\ U_m$	$0.35\ U_m/0.58\ U_m$
	220 kV	$U_m/\sqrt{3}$	0.10 $U_m(0.27U_m/0.46U_m)$	$0.75\ U_m$	0.35 k $U_m(0.35U_m/0.58\ U_m)$
	330–750 kV	$U_m/\sqrt{3}$	$0.10\ U_m$	$0.75\ U_m$	$0.35\ U_m$
Uneffective earthing	Unearthed	$1.10\ U_m$	$0.64\ U_m$	$1.38\ U_m$	$0.80\ U_m$
	Resonant earthing	U_m	$U_m/\sqrt{3}$	$1.25\ U_m$	$0.72\ U_m$
	Low resistor earthing	$0.80\ U_m$	$0.46\ U_m$	U_m	$U_m/\sqrt{3}$
	High resistor earthing	U_m	$U_m/\sqrt{3}$	$1.25\ U_m$	$U_m/\sqrt{3}$

Notes
1. The data above and below slash of 110 and 220 kV correspond to the conditions of system earthed or unearthed respectively
2. The data outside and inside of the bracket for 220 kV correspond to transformer neutral earthed through reactor and unearthed respectively
3 For 220 kV transformer neutral earthed through reactor and 330–750 kV transformer or High voltage shunt reactor earthed through reactor, when the ratio of the earthing reactor reactance and transformer or shunt reactor zero sequence reactance is n, then $k = 3n/(1 + 3n)$
4. This table is not applicable for to neutral unearthed of 110 and 220 kV transformers and insulation level lower than the values listed in GB 50064-2014 Table 6.4.6-3

Table 6 Typical temporary overvoltage

Neutral earthing type	Uneffective earthing (includes resistor and resonant earthing) kV			Solid earthing		
System nominal voltage (kV)	3–20	35–66	110–220	330–750 kV		
					Bus	Overhead Line
Temporary overvoltage (kV)	$1.1U_m$	U_m	$1.3\ \frac{U_m}{\sqrt{3}}$	$1.3\ \frac{U_m}{\sqrt{3}}$		$1.4\ \frac{U_m}{\sqrt{3}}$

- Fault cleared in 10 s, $U_r \leq 1.05U_{rG}$
- Fault cleared more than 10 s, $U_r \leq 1.25U_{rG}$

  U_{rG}: Generator rated voltage.

(3) MOA Continuous Operating Voltage (U_c):

U_c is approximately proportional to Ur. Normal $U_c = (0.76–0.8)\ U_r$, and details shall refer to the followings:

    ① Solid Earthing: $U_c \geq U_m//\sqrt{3}$
    ② Unsolid Earthing:

Fault cleared in 10 s: $U_c \geq U_m/\sqrt{3}$;
Fault cleared more than 10 s:

$U_c \geq U_m$ (35–66 kV)
$U_c \geq 1.1U_m$ (3–10 kV)

(4) Continuous Operating Voltage of surge arresters for generator protection shall be not less than generator rated voltage and 80% of MOA rated voltage.

5.1.3 Surge Arrester Parameter Selection as ABB Technical Guidance

"High Voltage Surge Arresters Buyer's Guide" [16] provides some guidance, details shall refer to Table 7.

ABB "Dimensioning, testing and application of metal oxide surge arresters in medium voltage networks" [17] provides the following guidance for medium voltage surge arrester selection.

For the arrester to meet the needs of the network system, two conditions are necessary to be fulfilled in the selection of the maximum continuous operating voltage Uc:

- Uc must be higher than the constant power frequency voltage at the arrester terminal.
- T × U_c must be higher than the expected temporary overvoltage at the arrester terminal. T is determined by the duration t of the temporary overvoltage. Thus in determining T, duration time is also to be taken into account.

Table 7 ABB surge arrester rated voltage selection

System earthing	Fault duration	System voltage U_m (kV)	Min. rated voltage U_r (kV)
Effective	≤1 s	≤100	≥0.8 × U_m
Effective	≤1 s	≥123	≥0.72 × U_m
Non-effective	≤10 s	≤170	≥0.91 × U_m ≥0.93 × U_m (EXLIM T)
Non-effective	≤2 h	≤170	≥1.11 × U_m
Non-effective	>2 h	≤170	≥1.25 × U_m

The table gives a minimum value of the arrester rated voltage (U_r). In each case, choose the next higher standard rating as given in the catalogue

Note Do not select a lower value of Ur than obtained as above unless the parameters are known more exactly; otherwise the arrester may be over-stressed by TOV

In selecting the arresters in a three-phase network, the location of the arrester plays the deciding role: between phase and earth, between transformer neutral and earth or between phases. The maximum operating voltage at the arrester terminal connection can be calculated with the help of the maximum voltage Um between phases. If this is not known, then Um should be replaced with the highest voltage of the system or the highest voltage for the electrical equipment.

In three-phase networks special attention must be paid to the temporary overvoltage UTOV. It occurs most frequently during earth faults. Its value is given by the method of neutral system earthing. Additionally the system management is of significance because it determines the duration t of the temporary overvoltage and with that it decides the value of $T_{(t)}$ for U_c.

$$U_c \geq U_{TOV}/T_{(t)}$$

(1) Networks with Earth Fault Compensation or with High-Ohmic Insulated Neutral

Under the conditions for earth-fault, the voltage increases at "healthy" phases to a maximum of U_m:

$$U_c \geq U_m$$

for an arrester between phase and earth.

The voltage at transformer neutral can reach a maximum of $U_m/\sqrt{3}$:

$$U_c \geq U_m/\sqrt{3}$$

for the arrester between transformer neutral and earth.

In every network there exists inductance and capacitance which produce oscillating circuits. If their resonant frequency is similar to that of the operating frequency, then the voltage between phase and earth can basically become higher than that of U_m in single pole earth faults. The system configuration must avoid the occurrence of such resonances. If this is not possible, then U_c of a corresponding magnitude should be chosen.

(2) Networks with High-resistance Insulated Neutral Systems and Automatic Earth Fault Clearing

Early cut-off of earth faults enables a reduction of U_c by the factor T. for example, the earth fault cutoff results after a maximum of t = 10 s, then, with the help of "Dimensioning, testing and application of metal oxide surge arresters in medium voltage networks", it follows that T = 1.26.

$$U_c \geq U_m/T$$

for an arrester between phase and earth.

$$U_c \geq U_m/(T \times \sqrt{3})$$

for the arrester between transformer neutral and earth.

(3)   Networks with Solidly Earthed Neutral Systems (Ce ≤ 1.4)

In this type of network, there are at least enough transformers in low-resistance neutral earthing, that during an earth fault the phase voltage never exceeds 1.4 p.u. in the entire system (earth fault factor Ce ≤ 1.4). Therefore is $U_{TOV} \leq 1.4 \times U_m /\sqrt{3}$. It can be assumed that the clearing time of the earth fault amounts to at the most t = 3 s. It follows for instance for the arrester that T = 1.28, and therefore:

$$U_c \geq 1.4U_m/(1.28 \times \sqrt{3}) = 1.1U_m/\sqrt{3}$$

for an arrester between phase and earth.

The voltage of the neutral of non-earth transformers reaches a maximum of UTOV $= 0.4 \times U_m$:

$$U_c \geq 0.4U_m/1.28 = 0.32U_m$$

for the arrester between transformer neutral and earth.

(4)   Networks with Low-resistance Neutral Earthing which do not have Uniformly Ce ≤ 1.4.

For arresters in the vicinity of neutral earth transformers, Uc can be chosen according to above section.

Care is required if the arrester is located just a few kilometers from the transformer, e.g. if there is a remote connection between an overhead line and a cable. If the ground is dried out or consists of rock, then it has a relatively high resistance. This can lead to a phase voltage at the location of the arrester which approaches Um. In this case it is recommended $U_c \geq U_m/T$ for an arrester between phase and earth.

Generally speaking, the earth fault monitoring would switch off the earth fault quickly (t < 3 s): therefore T = 1.28.

Under extremely poor earthing conditions (e.g. in desert regions), only a slight fault current flows in the case of a remote earth fault. If this is not caught by monitoring, switching off will not take place. The arresters in the vicinity of the earth fault are then loaded for a long period of time with Um. In such cases it is advisable to choose $U_C \geq U_m$.

For keeping in mind: If, as in the above described network, the arrester is located at a transformer with a low-ohmic neutral earthing, then $U_c \geq 1.4U_m/(1.28 \times \sqrt{3})$ is permissible. It is recommended that the earth connections of the arresters have a galvanic connection to the transformer tank and these connections should be kept as short as possible.

(5)   Low-resistance Neutral Earthing Networks and C > 1.4

This concerns networks which are earthed with an impedance so that the fault current is limited, for example, to 2 kA. In the case of an earth fault the voltage increases for a "healthy" phase to U_m.

With a resistive earthed neutral the voltage can be 5% higher than U_m. If the clearing time of the earth fault does not exceed t = 10 s, then results T = 1.26 (for the MWK):

$$U_c \geq 1.05U_m/T = 0.83U_m$$

5.2   Surge Arrester Insulation Coordination

International oil field surface engineering power system projects are often located in different countries and regions, with different voltage levels and system earthing types. Therefore, it is necessary to analyze and judge the insulation coordination to ensure insulation safety.

Technical specifications for equipment insulation and surge arrester residual voltage are as followings:

Lightning withstand of electrical equipment internal insulation:

$$U_{e.l.i} \geq K_16U_{l.p}$$

K_{16}—Coordination factor for electrical equipment internal insulation lightning withstand, when MOA installed nearby the protected equipment, $K_{16} = 1.25$, other condition, $K_{16} = 1.4$.

$U_{l.p}$—Surge arrester lightning residual voltage.

The following is a simple illustration to check whether a typical domestic standard 35 kV lightning arrester can be directly adopted in international 33 kV direct earthing system:

(1)   China 35 kV systems witch gear ($U_m = 40.5$ kV) typical lightning withstand voltage is 185 kV. Normal used surge arresters are 51/120-5-C; 51/134-5-C; 51/120-5-H; 51/134-5-H

(2)   Surge arrester lightning residual 134 kV

(3)   Coordination factor 185/134 = 1.38

(4)   International project 33 kV switchgear ($U_m = 36$ kV) typical lightning withstand voltage is 170 kV, when use China typical surge arrester in international project, the coordination factor is 170/134 = 1.27 > 1.25.

International project coordination factor can refer to Table 8.

Table 8 International coordination factors [18]

	400 kV	275 kV	132 kV	33 kV
Switching impulse withstand voltage (kVp)	1050	850		
Protection level (kV) (IEC)	$\frac{1050}{1.25} = 840$	$\frac{850}{1.25} = 680$	$\frac{550}{1.25} = 440$	$\frac{170}{1.25} = 136$
Lighting impulse withstand voltage (kVp)	1425	1050	650	170
Protection level (kV) (NGC)	$\frac{1425}{1.4} = 1020$	$\frac{1050}{1.4} = 750$	$\frac{650}{1.4} = 464$	$\frac{170}{1.4} = 121$
Maximum continuous voltage (kV)	$\frac{400 \times 1.23}{\sqrt{3}} = 285$	$\frac{275 \times 1.23}{\sqrt{3}} = 195$	$\frac{132 \times 1.23}{\sqrt{3}} = 94$	$\frac{33 \times 1.23}{\sqrt{3}} = 23$
Rated surge arrester voltage (kV) ≥ (NGC)	$\frac{400 \times 1.58}{\sqrt{3}} = 365$	$\frac{275 \times 1.58}{\sqrt{3}} = 250$	$\frac{132 \times 1.58}{\sqrt{3}} = 120$	$\frac{33 \times 2.2}{\sqrt{3}} = 42$
Energy level class ≥ normally	3 4	3 4	3 4	3 4
Nominal discharge current	10–20	10–50	10	

5.3 SPD

The Surge Protective Device (SPD) is an International Electrotechnical Commission (IEC) definition of a Surge protector. It is a Device used to limit transient overvoltage and pilot surges. When the electrical circuit or communication circuit is interfered by the external interferences which produce a peak voltage or current, SPD can conduct shunt in an instant to avoid damage to equipment.

Domestic Electrical Industry standard, DL 5408 [19], specified where shall install SPDs and which type of SPD shall be applied. 5.3.1 and 5.3.2 specified where should be provide with SPDs of power plant and substation as per DL 5408.

5.3.1 SPD Installations in Power Plant

Large and medium scale power plant LV (220/380 V) system should be provided in the following power supplies.

– 220/380 V Power Center (PC) bus incomers and emergency bus section.
– Electronic information equipment power incoming.
– AC uninterruptable power supply (UPS) 220/380 V incomer.

Impulse current and nominal discharge current should be Class A.

Small scale power plant LV (220/380 V) system should be provided in the following power supplies.

– 220/380 V Power Center (PC) bus incomers.
– Electronic information equipment power incoming.
– C uninterruptable power supply (UPS) 220/380 V incomer.

Impulse current and nominal discharge current should be Class B.

5.3.2 SPD Installations in Substation

35 kV and above substation LV (220/380 V) system should be provided in the following power supplies.

- Substation auxiliary power supply transformer LV bus incomer
- Incomer of AC uninterruptable power supply (UPS) 220/380 V for monitoring computer

Impulse current and nominal discharge current should be Class B.

35 kV below substation LV (220/380 V) system should be provided in the following power supplies.

- Substation auxiliary power supply LV bus incomer
- Substation LV distribution bus

Impulse current and nominal discharge current should be Class C.

SPD Classification, discharge current and withstand voltage shall refer to Tables 9 and 10.

Table 9 Power supply circuit SPD discharge current [19]

Protection class	Boundary between LPZ0 and LPZ1	Boundary between LPZ1 and LPZ2, LPZ2 and LPZ3			DC norminal impulse current (kA)
	Class I impulse current (kA)	Class II impulse current (kA)	Class III impulse current (kA)	Class IV impulse current (kA)	8/20 μs
	10/350 μs	8/20 μs	8/20 μs	8/20 μs	
Class A	≥20	≥40	≥20	≥10	≥10
Class B	≥15	≥40	≥20		In the DC distribution system, the SPD with nominal discharge current ≥10 kA is selected according to the line length and working voltage
Class C	≥12.5	≥20			
Class D	≥12.5	≥10			

Note The outer packing material of SPD shall be flame retardant material

Table 10   Indoor 220/380 V distribution system equipment insulation level [20]

Installation location	Power source	Distribution circuit and equipment in final branch	Appliances	Special protected equipment
Impulse voltage category	IV	III	II	I
Rated withstand voltage U_w (kV)	6	4	2.5	1.5

Notes

1. Category I—contains electronic circuit devices, such as computer, with the electronic process control equipment
2. Category II—such as household appliances and similar load
3 Category III—such as panel, circuit breaker, including line, busbar, junction box, switch, socket device such as a fixed wiring system, as well as the application in industrial equipment and permanently connected to the fixed device
4 Category IV—such as electric measuring instrument, a line over-current protection device, filter

6   Substation Safety Clearance

Safety Clearance is a critical issue in electrical engineering of indoor and outdoor substations. There are different specifications for safety clearance in GB (China National Standard), IEEE and IEC standards. For international project, suitable standards should be followed and ITB together with local law shall be the first consideration in common engineering practice.

As per actual practices, GB's technical specifications are in more detailed in electrical equipment layout than that of in internal standards. Although some specific clearances in international standards are different with GB's specification, so in international substation design, comparison and analysis of safety clearance shall be checked firstly.

For fireproof distance, international standards are normally stricter than GB, so this part shall be more attention in actual engineering.

This section summarizes substation safety clearance and fire proof distance of indoor & outdoor substation in main international standards and China National standards for engineering reference.

6.1   Safety Clearance of Outdoor Switchyard

6.1.1   Safety Clearance Requirement in GB 50060 [21]

Safety clearance of outdoor switchyard shall not be less than the values listed in Table 11. Which the lowest part of the external insulator of electrical equipment is less than 2500 mm from the ground, a fixed barrier shall be installed.

Which clearance of outdoor switchyard shall be complied as Figs. 6, 7, 8 and

Table 11 Safety clearance of outdoor switchyard [20]

Symbol	Application scope	System nominal voltage (kV)					
		3–10	15–20	35	66	110 J	110
A1	1. Between the live part and the ground part 2. The upward extension line of the mesh hood is between 2.5 m from the ground and the live part above the hood	200	300	400	650	900	1000
A2	1. Between live parts of different phases 2. Between live leads on both sides of the break of the circuit breaker and disconnect	200	300	400	650	1000	1100
B1	1. During equipment transportation, the equipment outer surface and the live part without blocking 2. Crossover between the different live parts of maintenance(power switching off) in different times 3. Palisade block between the insulator and the live part 4. Between the live part and the ground part during live operation	950	1050	1150	1400	1650	1750
B2	Mesh blocking to live parts	300	400	500	750	1000	1100
C	1 Between bare conductor (without block) and ground 2. Bare conductor (without block) and the top of the building and structure	2700	2800	2900	3100	3400	3500
D	1. Parallel and different blackout maintenance between unblocked live parts 2. Between the live part and the edge part of the building and structure	2200	2300	2400	2600	2900	3000

Notes
(1) 110 J refers an effective neutral earthing system
(2) Where the altitude exceeds 1000 m, 'A' should be corrected
(3) The values listed in this table are not applicable for the manufacturer's complete package of power distribution units
(4) During live work, the B1 value can be added 750 mm to the A2 value between live parts of different phases or intersecting circuits

Tables 11 and 12.

6.1.2 Safety Clearance Requirement in IEC 61936-1 [22]

IEC 61936-1 specified the following clearances:

- Minimum phase-to-earth and phase-to-phase clearance (N).
- Protective barrier clearances (B1 & B2).

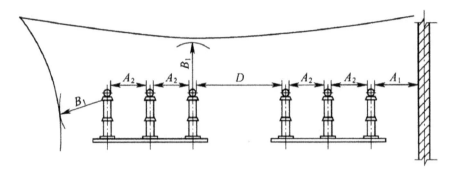

Fig. 6 Outdoor A1, A2, B1, D

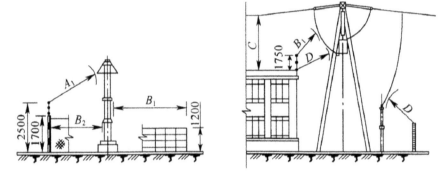

Fig. 7 Outdoor A1, B1, B2, C, D

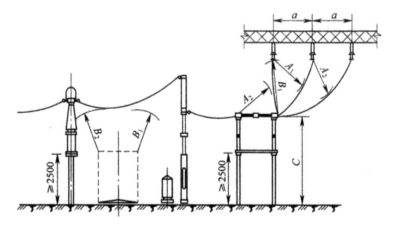

Fig. 8 Outdoor A1, A2, B1, B2, C. *Notes* (1) A is the distance between live parts of different phases. (2) When soft conductor is used for outdoor power distribution device, the minimum safe clean distance between live part and ground part and live part of different phases under different conditions should be checked according to Table 11, and the maximum value should be adopted

Table 12 Minimum clearance of live parts to earth and phase to phase [21]

Condition	Environment	Design wind speed, m/s	A	System nominal voltage, kV			
				35	66	110 J	110
Lightning overvoltage	Lightning overvoltage and wind swing	10	A1	400	650	900	1000
			A2	400	650	1000	1100
Power frequency overvoltage	(1) Max. working voltage, short circuit and wind swag (at 10 m/s) (2) Max. working voltage, wind swag (at Max. design wind speed)	10 or Max. design wind speed	A1	150	300	300	450
			A2	150	300	500	500

- Protective obstacle clearances (O2).
- Boundary clearances (C & E).
- Minimum height over access area (H).
- Clearances to buildings (Dv B2 N).
- Transport routes (T).
- Electrical installations on mast, pole and tower (H').

The following sections specify details of above clearance.

(1) Minimum phase-to-earth and phase-to-phase clearance (N)

(1) $U_n = 66$ kV, IEC61936-1 minimum phase-to-earth N = 630 mm (GB50060 A1 = 650 mm); IEC61936-1 minimum phase-to-phase N = 630 mm (GB50060 A2 = 650 mm).

(2) $U_n = 110$ kV, IEC61936-1 minimum phase-to-earth N = 900 mm (Power frequency withstand voltage 185 kV, Lightning withstand voltage 450 kV);GB 50060 A1 = 900 mm for solid earthing system); IEC61936-1 minimum phase-to-phase N = 1100 mm (Power frequency withstand voltage 230 kV, Lightning withstand voltage 550 kV GB50060 A1 = 900 mm for solid earthing system).

More clearances specified in IEC 61936-1 shall refer to Table 13.

(2) Protective barrier clearances (B1 & B2)

Within an installation, the following minimum protective clearances shall be maintained between live parts and the internal surface of any protective barrier:

- for solid walls, without openings, with a minimum height of 1800 mm, the minimum protective barrier clearance is B1 = N;

Table 13 Minimum clearance of phase to phase and phase to ground

Highest voltage for installation	Rated short-duration power frequency withstand voltage	Rated lightning impulse with stand voltage	Minimum phase-to-earth and phase-to-phase clearance N	
U_m r.m.s.	U_d r.m.s.	U_p 1, 2/50 μs (peak value)	Indoor installations	Outdoor installations
kV	kV	kV	mm	mm
3.6	10	20	60	120
		40	60	120
7.2	20	40	60	120
		60	90	120
12	28	60	90	150
		75	120	150
		95	160	160
17.5	38	75	120	160
		95	160	160
24	50	95	160	
		125	220	
		145	270	
36	70	145	270	
		170	320	
52	95	250	480	
72.5	140	325	630	
123	185[b]	450[b]	900	
	230	550	1100	
145	185[b]	450[b]	900	
	230	550	1100	
	275	650	1300	

– for wire meshes, screens or solid walls with openings, with a minimum height of 1800 mm and a degree of protection of IPXXB (see IEC 60529), the minimum protective barrier clearance is B2 = N + 80 mm.

Note The degree IPXXB ensures protection against access to hazardous parts with fingers.

For non-rigid protective barriers and wire meshes, the clearance values shall be increased to take into account any possible displacement of the protective barrier or mesh.

(3) Protective obstacle clearances (O2)

Within installations the following minimum clearance shall be maintained from live parts to the internal surface of any protective obstacle:

- for solid walls or screens less than 1800 mm high, and for rails, chains or ropes, the minimum protective obstacle clearance is O2 = N + 300 mm (minimum 600 mm);
- for chains or ropes, the values shall be increased to take into account the sag.

Where appropriate, protective obstacles shall be fitted at a minimum height of 1200 mm and a maximum height of 1400 mm.

Note Rails, chains and ropes are not acceptable in certain countries.

(4) Boundary clearances (C & E)

The external fence of outdoor installations of open design shall have the following minimum boundary clearances:

- solid walls: C = N + 1000 mm;
- wire mesh/screens: E = N + 1500 mm.

(5) Minimum height over access area (H)

The minimum height of live parts above surfaces or platforms where only pedestrian access is permitted shall be as follows:

- for live parts without protective facilities, a minimum height H = N + 2250 mm (minimum 2500 mm) shall be maintained. The height H refers to the maximum conductor sag;
- the lowest part of any insulation, for example the upper edge of metallic insulator bases, shall be not less than 2250 mm above accessible surfaces unless other suitable measures to prevent access are provided.

Where the reduction of safety distances due to the effect of snow on accessible surfaces needs to be considered, the values given above shall be increased.

(6) Clearances to buildings (D_v B2 N).

Where bare conductors cross buildings which are located within closed electrical operating areas, the following clearances to the roof shall be maintained at maximum sag:

- H sag value if the roof is accessible when the conductors are live;
- N + 500 mm where the roof cannot be accessed when the conductors are live;
- O_2 in lateral direction from the end of the roof if the roof is accessible when the conductors are live.

Where bare conductors approach buildings which are located within closed electrical operating areas, the following clearances shall be maintained, allowing for the maximum sag/swing in the case of stranded conductors:

- outer wall with unscreened windows: Minimum clearance given by D_v;
- outer wall with screened windows: Protective barrier clearances B2;
- outer wall without windows: N.

(7) Transport routes (T)

Transport routes, their load capacity, height and width shall be adequate for movements of anticipated transport units and shall be agreed upon between the supplier and the user.

Within closed electrical operating areas, the passage of vehicles or other mobile equipment beneath or in proximity to live parts (without protective measures) is permitted, provided the following conditions are met:

– the vehicle, with open doors, and its load does not infringe the danger zone: Minimum protective clearance for vehicles T = N + 100 (minimum 500 mm);
– the minimum height, H, of live parts above accessible areas is maintained.

Under these circumstances, personnel may remain in vehicles or mobile equipment only if there are adequate protective measures on the vehicle or mobile equipment, for example the cab roof, to ensure that the danger zone defined above cannot be infringed.

For the lateral clearances between transport units and live parts, similar principles apply.

(8) Installations in tower or pole/structure

The minimum height H′ of live parts above surfaces accessible to the general public shall be H′ = 4300 mm for rated voltages Um up to 52 kV;
• H′ = N + 4500 mm (minimum 6000 mm) for rated voltages Um above 52 kV;
Where N is the minimum clearance.

Where the reduction of safety distances due to the effect of snow on accessible surfaces needs to be considered, the values given above shall be increased.

Isolating equipment and fuses shall be arranged so that they can be operated without danger.

Isolating equipment accessible to the general public shall be capable of being locked. The operating rods shall be compliant with the relevant standard.

Safe phase-to-phase connection and earthing of the overhead line shall be possible.

Figures 9, 10, 11, 12, 13, 14, and 15 are reference to above safety distance.

6.2 Safety Clearance of Indoor Substation

6.2.1 Safety Clearance Requirement in DL 5352 [23]

Minimum width of all paths for metal enclosed switchgears in the substation should be in accordance with the provisions of Table 14.

For the indoor oil-immersed transformer, minimum clearance between transformer enclosure and walls shall refer to Table 15. When the indoor oil-immersed transformer needs on-site maintenance, the transformer room height can be increased

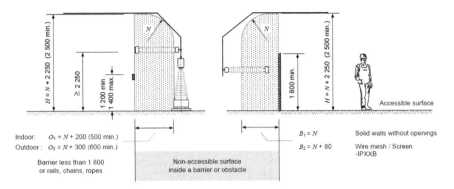

Fig. 9 Protection against direct contact by protective barriers/protective obstacles within closed electrical operating areas. *Note* N—Minimum clearance, O—Obstacle clearance, B—Barrier clearance

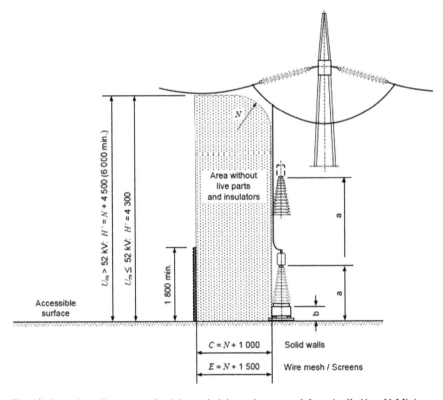

Fig. 10 Boundary distances and minimum height at the external fence/wall. *Note* N Minimum clearance; H' Minimum clearance of live parts above accessible surface at the external fence; (a) If this distance to live parts is less than H, protection by barriers or obstacles shall be provided; (b) If this distance is smaller than 2250 mm, protection by barriers or obstacles shall be provided

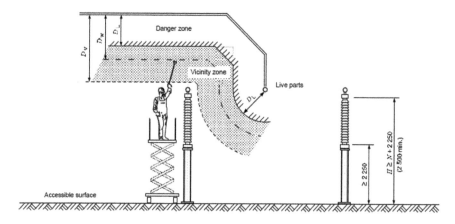

Fig. 11 Minimum heights and working clearances within closed electrical operating area. *Note* D_L—N; D_V—N + 1000 for Un $\leq$110 kV; D_V—N + 2000 for Un >110 kV; D_W—Working clearance according to national standards or regulations; N—Minimum clearance; H—Minimum height

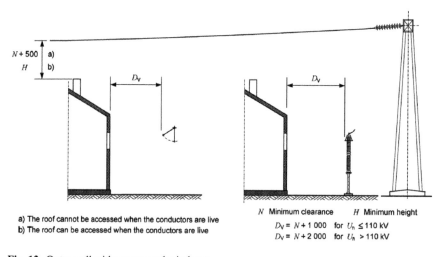

a) The roof cannot be accessed when the conductors are live
b) The roof can be accessed when the conductors are live

N Minimum clearance H Minimum height
$D_V = N + 1\,000$ for $U_h \leq 110$ kV
$D_V = N + 2\,000$ for $U_h > 110$ kV

Fig. 12 Outer wall with unscreened windows

by 700 mm according to the minimum height required by the hanging core, and the width can be determined by adding 800 mm on both sides of the transformer.

For the indoor without enclosure dry transformer, clearance between transformer external contour and walls shall be not less than 600 mm, and clearance between transformers shall be not less than 1000 mm. Patrol requirement shall be also ensured. For the indoor enclosed dry transformer, clearances are not limited by the above clearances for indoor without enclosure dry transformer.

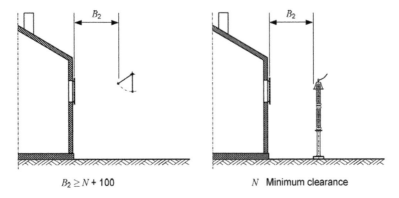

$B_2 \geq N + 100$ N Minimum clearance

Fig. 13 Outer wall with screened windows

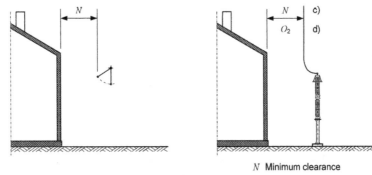

N Minimum clearance

c) N if the roof is non-accessible when the conductors are live
d) $O_2 \geq N + 300$ (600 min.) if the roof is accessible when the conductors are live

Fig. 14 Outer wall without windows

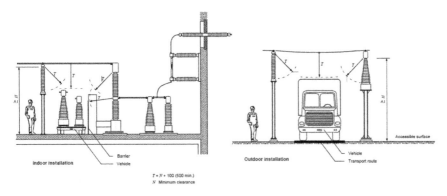

$T = N + 100$ (500 min.)
N Minimum clearance

Fig. 15 Minimum approach distance for transport

Table 14 Minimum width for indoor metal enclosed switchgears (mm)

Access type arrangement	Maintenance access	Operation access	
		fix equipment	Truck type equipment
Single row	800	1500	Truck length + 1200
Double rows	1000	2000	Two truck length + 900

Notes

(1) The width of the access can be reduced by 200 mm at the individual protrusions of the wall columns of the building

(2) When the truck cabinet does not need on-site maintenance, its access width can be appropriately reduced

(3) When fixed type cabinet is arranged against the wall, the distance from the cabinet to the wall should be 50 mm

(4) When 35 kV switchgear is adopted, the access behind the cabinet should not be less than 1000 mm

Table 15 Minimum clearance between oil-immersed transformer and transformer room wall (mm)

Transformer rating	1000 kVA and below	1250 kVA and above
Distance between transformer enclosure to rear wall and side wall	600	800
Distance between transformer enclosure and door	800	1000

6.2.2 Safety Clearance Requirement in GB 50054 [24]

Low voltage switchgear room minimum path width shall refer to Table 16.

6.2.3 Safety Clearance Requirement in DL 5153 [25]

Minimum path width of low voltage switchgear shall refer to Table 17.

6.2.4 Safety Clearance Requirement in NFPA 70 [26]

Minimum depth of clear working space at electrical equipment in NFPA 70 shall refer to Table 19.

Table 19 is suitable for the following three conditions:

(1)   Condition 1—Exposed live parts on one side of the working space and no live or grounded parts on the other side of the working space, or exposed live parts on both sides of the working space that are effectively guarded by insulating materials.

Table 16 Low voltage switchgear room minimum path width (m)

SWGR		Single row			Double row (face-to-face)			Double row (back-to-back)			Multiple rows in the same direction			Side walk way
		Front	Rear		Front	Rear		Front	Rear		Panel to panel	Front and rear to wall		
			Maintenance	Operation		Maintenance	Operation		Maintenance	Operation		Front	Rear	
Fix	Un-restricted	1.5	1.1	1.2	2.1	1	1.2	1.5	1.5	2.0	2.0	1.5	1.0	1.0
	Restricted	1.3	0.8	1.2	1.8	0.8	1.2	1.3	1.3	2.0	1.8	1.3	0.8	0.8
Truck	Un-restricted	1.8	1.0	1.2	2.3	1.0	1.2	1.8	1.0	2.0	2.3	1.8	1.0	1.0
	Restricted	1.6	0.8	1.2	2.1	0.8	1.2	1.6	0.8	2.0	2.1	1.6	0.8	0.8

Notes

(1) Restricted means restricted by the building plan, columns in the access and other local protrusions

(2) Rear operation access for operator at rear of the switchgear

(3) The width of the front access of the switchgears in back-to-back layout can be determined according to the width of the double row (back-to-back) layout in this table.

(4) The minimum width of the front and rear access of control panel, control cabinet/panel and floor mounted power distribution board can be determined according to this table.

(5) The width of the operation access in front of the wall-mounted distribution board should not be less than 1 m

For floor installed distribution board, the distribution board bottom shall be minimum 50 mm above the finished ground indoor and minimum 200 mm above the finished ground outdoor

Where the high voltage and low voltage distribution equipment are located in the same room and there is a exposed bus on the top of one side of the switchgear, the clearance between high-voltage and low-voltage distribution equipment should not be less than 2 m

Where low voltage switchgears arranged in row, and the length is more than 6 m, there shall be two exits at rear of the switchgear row and arranged at both ends. Where the distance between two exits exceeds 15 m, additional exit shall be added in between

Table 17 Minimum path width of low voltage switchgear (front and rear path, m)

SWGR. type		Single row – Front	Single row – Rear Maintenance	Single row – Rear Operation	Double row (face-to-face) – Front	Double row (face-to-face) – Rear Maintenance	Double row (face-to-face) – Rear Operation	Double row (back-to-back) – Front	Double row (back-to-back) – Rear Maintenance	Double row (back-to-back) – Rear Operation	Multiple rows same direction – Panel to panel	Front and rear to wall – Front	Front and rear to wall – Rear
Fix	Un-restricted	1.5	1.0	1.2	2.0	1.0	1.2	1.5	1.5	2.0	2.0	1.5	1.0
Fix	Restricted	1.3	0.8	1.2	1.8	0.8	1.2	1.3	1.3	2.0	2.0	1.3	0.8
Truck	Un-restricted	1.8	1.0	1.2	2.3	1.0	1.2	1.8	1.0	2.0	2.3	1.8	1.0
Truck	Restricted	1.6	0.8	1.2	2.0	0.8	1.2	1.6	0.8	2.0	2.0	1.6	0.8

Notes

(1) Restricted means restricted by the building plane, columns in the passage and other local protrusions

(2) The minimum width of the access before and after the control panel and cabinet can be implemented or reduced appropriately according to the provisions of this table

(3) Rear operation access means the access for the rear operation for the switchgear in working

(4) When the cabling position is in the front of the panel and panel cabinet is arranged against the wall, then min. 200 mm space should be provided and bottom incoming is preferred

The clearance between indoor live part and other part shall meet the following requirements in the low voltage plant substation

When the height of the bare conductive part in the rear panel is lower than 2.3 m, protection meshes shall be provided and the path height shall be not less than 1.9 m after adding protection meshes

When the height of the bare conductive part in the rear panel is lower than 2.5 m, protection meshes shall be provided and the path height shall be not less than 2.2 m after adding protection meshes

Medium voltage plant substation clearances shall refer to Table 18

Table 18 Medium voltage plant substation clearances (mm)

SWGR. type		Single row			Double row (face-to-face)			Double row (back-to-back)			Multiple rows in the same direction		
		Front	Rear		Front	Rear		Front	Rear		Front	Rear	
			Maintenance	Operation		Maintenance	Operation		Maintenance	Operation		Maintenance	Operation
Fix	Un-restricted	1500	1000	1200	2000	1000	1200	1500	1500	2000	2000	1500	1000
	Restricted	1300	800	1200	1800	800	1200	1300	1300	2000	2000	1300	800
Withdraw	Unrestricted	1800	1000	1200	2300	1000	1200	1800	1000	2000	2300	1800	1000
	Restricted	1600	800	1200	2000	800	1200	1600	800	2000	2000	1600	800

Notes
(1) The dimensions in the table are calculated from the screen surface of the commonly used switch cabinet (that is, the protrusions have been included in the dimensions in the table)
(2) The dimensions of the operation and maintenance accesses listed in the table are allowed to be reduced by 200 mm in individual protruding parts of the building

Table 19 Minimum depth of clear working space at electrical equipment

Nominal voltage to ground	Minimum clear distance		
	Condition 1	Condition 2	Condition 3
1001–2500 V	900 mm (3 ft)	1.2 m (4 ft)	1.5 m (5 ft)
2501–9000 V	1.2 m (4 ft)	1.5 m (5 ft)	1.8 m (6 ft)
9001–25000 V	1.5 m (5 ft)	1.8 m (6 ft)	2.8 m (9 ft)
25001 V–75 kV	1.8 m (6 ft)	2.5 m (8 ft)	3.0 m (10 ft)
Above 75 kV	2.5 m (8 ft)	3.0 m (10 ft)	3.7 m (12 ft)

(2) Condition 2—Exposed live parts on one side of the working space and grounded parts on the other side of the working space. Concrete, brick, or tile walls shall be considered as grounded.

(3) Condition 3—Exposed live parts on both sides of the working space.

6.2.5 Safety Clearance Requirement in IEEE C2 [27]

IEEE C2 Rule 180B specified the following requirement for equipment layout:

1. Switchgear shall not be located within 7.6 m (25 ft) horizontally indoors or 3.0 m (10 ft) outdoors of storage containers, vessels, utilization equipment, or devices containing flammable liquids or gases.
    EXCEPTION: If an intervening barrier, designed to mitigate the potential effects of flammable liquids or gases, is installed, the distances listed above do not apply.
2. Enclosed switchgear rooms shall have at least two means of exit, one at each extreme of the area, not necessary in opposite walls. Doors shall swing out and be equipped with panic bars, pressure plates, or other device that are normally latched but open under simple pressure.
    EXCEPTION: One door may be accepted when required by physical limitations if means are provided for unhampered exit during emergencies.
3. Space Shall Be Maintained in Front the Switchgear to Allow Breakers to Be Removed and Turned Without Obstruction
4. Space shall be maintained in the rear of the switchgear to allow for door opening to at least 90 degrees open, or a minimum of 900 mm (3ft) without obstruction when removable panels are adopted.

6.3 Transformer Fire Proof Distance

6.3.1 Transformer Fire Proof Distance as GB50060 [20]

When indoor open type electrical equipment contains oil and the rated voltage is 35 kV, the equipment shall be installed in the space with partition walls (boards) on both sides.

When indoor open type electrical equipment contains oil and the rated voltage is 66–110 kV, the equipment shall be installed in the bay with explosion-proof partition wall.

Indoor oil-immersed power transformers with a total oil content of more than 100 kg shall be installed in an independent room with fire extinguishing facilities.

When the oil quantity of single electrical equipment in the room is more than 100 kg, oil storage facilities or oil retaining facilities shall be provided. The capacity of oil retaining facilities shall be designed to accommodate 20% of the oil, and the facilities shall be provided for discharging the oil to a safe place. When the above requirements can not be met, an oil storage facility shall capable of holding 100% oil. The inner diameter of the pipe for oil drain shall not be less than 150 mm, and the pipe inlet shall be equipped with an iron-grid filter.

When the oil quantity of single outdoor electrical equipment is more than 1000 kg, oil storage or oil retaining facilities should be provided. When a storage or containment facility containing 20% of the oil been installed, facilities shall be installed to discharge the oil to a safe place and shall not cause pollution. When the above requirements cannot be met, oil storage or oil retaining facilities that can hold 100% oil shall be installed. The oil storage and oil retaining facilities shall be larger than 1000 mm on each side of the external profile of the equipment and 100 mm above the ground on all sides. Pebble layer shall be laid in the oil storage facility. The thickness of pebble layer shall not be less than 250 mm, and the diameter of pebble shall be 50–80 mm. When the total accident oil storage tank with oil–water separation measures is set, the oil storage tank capacity should be determined by 60% of the maximum capacity of one transformer oil tank.

The minimum clearance between outdoor oil-immersed transformers with an oil content of 2500 kg or more shall comply with the provisions of Table 20.

The minimum clearance for 220 and 330 kV is 10 m [28]. The minimum clearance for 550 kV is 15 m [28].

When the fire proof distance between outdoor oil-immersed transformers with an oil volume of 2500 kg or more cannot meet the requirements of Table 20, firewalls shall be provided. The fire endurance of firewall should not be less than 4 h. The height of the firewall should be higher than the transformer oil conservator, and its length should be greater than 1 m on both sides of the transformer oil storage pool.

The fire proof distance between outdoor oil-immersed transformers or reactors with an oil content of 250 kg or more and oil-filled electrical equipment with an oil content of 600–2500 kg in the same circuit/bay shall not be less than 5 m.

In places with higher fire prevention requirements, non-oil insulated electrical equipment should be used if possible.

Rated voltage (kV)	Distance (m)
35 and below	5
66	6
110	8

Table 20 Minimum distance between outdoor oil-immersed transformers

6.3.2 Transformer Fire Proof Distance as IEC 61936-1 [22]

China national standard of oil-immersed transformer fire proof separation is based on voltage level, but in general international standard separation is based on the properties and total weight of transformer oil. According to IEC standard, the clearance is divided into four classes as per the total weight of insulating oil, IEEE is with two classes and the national standard is with three classes as per the voltage level. If transformer outside profile spacing is less than the specific distance, firewall shall be required for safety protection.

Maximum fire proof separation is 8 m as per China national in common conditions, and maximum fire proof separation is 15/30 m for oil insulated transformers with 45000 L oil as per IEC 61936-1, maximum fire proof separation is 15.2 m for mineral-oil-insulated transformers with 45000 L oil as per IEEE 979 [29].

Particularly, China national standard fire resistance durance time requirement is more stringent than the requirement in IEC and IEEE based on the above (Fig. 16).

Fire proof separation for outdoor transformer to transformer and transformer to building shall refer to Table 21.

Meaning of G1 and G2 which indicated in Tables 21 and 22 shall refer to Figs. 17 and 18.

Minimum requirements for the installation of indoor transformers are given in Table 22.

6.3.3 Transformer Fire Proof Distance as IEEE 979

1 *Equipment to equipment*:

Individual space of mineral-oil-insulated equipment should be separated from the anticipated flame fronts of one another by the distances given in Table 23. Separation distances to adjacent equipment should be measured from the edge of the postulated flame front to the nearest mineral-oil-filled component of the adjacent equipment.

2 *Equipment to buildings*:

Noncombustible or limited combustible buildings should be separated from adjacent mineral-oil-insulated equipment containment area(s) by a 2 h rated firewall or the separation values in Table 23.

The main differences between IEEE 979 and China National standard and IEC 61936-1 are as follows:

- The fire spacing stipulated by IEEE 979 is generally larger than the national standard;
- The maximum fire spacing prescribed by IEEE 979 is 15.2 m, compared to that prescribed by IEC 61936-1, which is 30 m.
- IEEE 979 stipulates that if the fireproof spacing does not meet the requirements of Table 5.3-4, firewalls should be set between devices, and the fireproof time

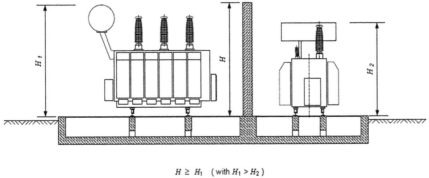

$$H \geq H_1 \quad (\text{with } H_1 > H_2)$$
$$L \geq B_2 \quad (\text{with } B_2 > B_1)$$

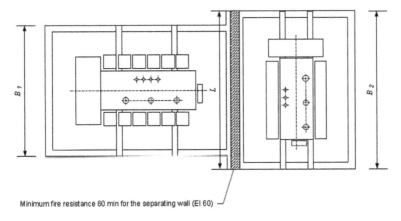

Minimum fire resistance 60 min for the separating wall (EI 60)

Fig. 16 Separating walls between transformers

of firewalls should not be less than 2 h, which is greatly different from the 4 h stipulated by the national standard.

7 Miscellaneous Safety Points

7.1 Safety Interlocks for Switchgears

There are five mandatory interlock provisions for switchgear in China to ensure the safety of human and equipment. These mandatory interlocks are designed to avoid of destroy of mal-operation and are very important safe production. On the contrary, there are no such kind of mandatory interlock requirement for switchgear interlocks

Table 21 Guide values for outdoor transformer clearances [22]

Transformer type	Liquid volume I	Clearance G1 to other transformers or building surface of non-combustible material, m	Clearance G2 to building surface of combustible material, m
Oil insula ted transformers (O)	$1000 < \cdots < 2000$	3	7.5
	$2000 \leq \cdots < 20000$	5	10
	$20000 \leq \cdots < 45000$	10	20
	≥ 45000	15	30
Less flammable liquid insulated transformers (K) without enhanced protection	$1000 < \cdots < 3800$	1.5	7.5
	≥ 3800	4.5	15

Less flammable liquid insulated transformers (k) with enhanced protection	Clearance G1 to building surface or adjacent transformers	
	Horizontal, m	Vertical, m
	0.9	1.5

Dry-type transformers (A)	Fire behavior class	Clearance G1 to building surface or adjacent transformers	
		Horizontal, m	Vertical, m
	F0	1.5	3.0
	F1	None	None

Note 1 Enhanced protection means
– Tank rupture strength
– Tank pressure relief
– Low-current fault protection
– High-current fault protection
For an example of enhanced protection, see Factory Mutual Global standard 3990, or equivalent
Note 2 Sufficient space should be allowed for periodic cleaning of resin-encapsulated transformer windings, in order to prevent possible electrical faults and fire hazard caused by deposited atmospheric pollution
Note 3 Non-combustible materials may be chosen in accordance to EN 13501-1

Table 22 Minimum requirements for the installation of indoor transformers

Transformer type	Class	Safeguards
Oil insulated transformers (O)	Liquid volume	
	≤1000 L	EI 60 respectively REI 60
	>1000 L	EI 90 respectively REI 90 or EI 60 respectively REI 60 and automatic sprinkler protection
Less flammable liquid insulated transformers (K)	Nominal power/max. voltage	
Without enhanced protection	(No restriction)	60 respectively REI 60 or automatic sprinkler protection
With enhanced protection	≤ 10 MVA and Um ≤ 38 kV	60 respectively REI 60 or separation distances 1.5 m horizontally and 3.0 m vertically
Dry-type transformer (A)	Fire behavior class	
	F0	EI 60 respectively REI 60 or separation distances 0.9 m horizontally and 1.5 m vertically
	F1	Non combustible walls

Note 1 REI represents the bearing system (wall) whereas EI represents the non-load bearing system (wall) where R is the load bearing capacity. E is the fire integrity. I is the thermal insulation and 60/90 refers to fire resistance duration in minutes

Note 2 Definitions of fire resistance are given in EN 13501-2

Note 3 Enhanced protection means
　Tank rupture strength
– Tank pressure relief
– Low-current fault protection
– High-current fault protection
For an example of enhanced protection, see Factory Mutual Global standard 3990 [30], or equivalent

Note 4 Sufficient space should be allowed for periodic cleaning of resin-encapsulated transformer windings, in order to prevent possible electrical faults and fire hazard caused by deposited atmospheric pollution

in other similar international project that is may be depending on local standards and specification, which the mandatory technical interlock should be advised and to be adopted in detail design.

7.1.1 Switchgear Interlock Requirement as Per GB 50060 [21]

Metal clad switchgear of 35 kV and below shall have the following functions:

– Prevent mis-opening and mis-closing of circuit breaker;
– Prevent onload open and close of isolation switch;
– Prevent put grounding wires/close earthing switch when main circuit still been energized;

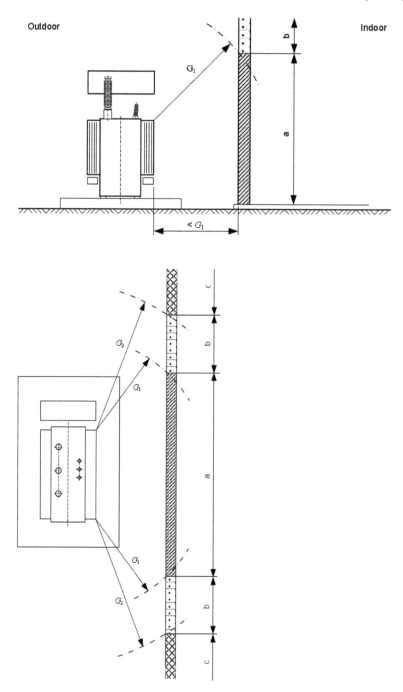

Fig. 17 Fire protection between transformer and building surface of non-combustible material

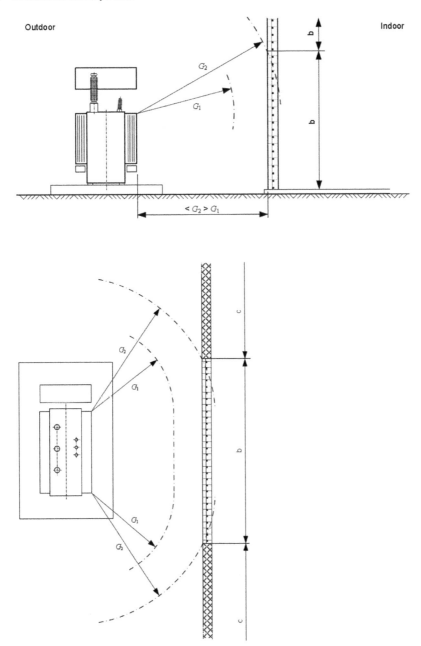

Fig. 18 Fire protection between transformer and building surface of combustible material. *Note* For Clearances G1 and G2, see Tables 21 and 22; Sector a The wall in this area shall be designed with a minimum fire resistance of 90 min (REI 90); Sector b The wall in this area shall be designed with non-combustible materials; Sector c No fire protection requirements. *Note* Due to the risk of vertical fire spread sector c applies only in the horizontal direction

Table 23 Separation
distances

Mineral oil volume, L (gal)	Separation distance, m (ft)
<1890 (500)[a]	
1890 to 18930 (500 to 5000)	7.6 (25)
>18930 (5000)	15.2 (50)

[a]Determining the type of physical separation to be used for mineral oil volumes less than 1890 L (500 gal) should be based on consideration of the following:
– Type and quantity of oil in the equipment
– Size of a postulated oil spill (surface area and depth)
– Construction of adjacent structures
– Rating and bushing type
– Fire-suppression systems provided
– Protection clearing time

– Prevent close circuit breaker (disconnector) with grounding wire/earthing switch is still in working status;
– Prevent mis-entry of live/energized switchgear.

7.1.2 Switchgear Interlock Requirement as Per IEC 62271-200 [30]

Interlocks between different components of the equipment are provided for reasons of protection and for convenience of operation. The following provisions are mandatory for main circuits.

(a) *Metal-enclosed switchgear and controlgear with removable parts*

The withdrawal or engagement of a circuit-breaker, switch or contactor shall be prevented unless it is in the open position.

The operation of a circuit-breaker, switch or contactor shall be prevented unless it is only in the service, disconnected, removed, test or earthing position.

The interlock shall prevent the closing of the circuit-breaker, switch or contactor in the service position unless any auxiliary circuits associated with the automatic opening of these devices are connected. Conversely, it shall prevent the disconnection of the auxiliary circuits with the circuit-breaker closed in the service position.

(b) *Metal-enclosed switchgear and controlgear provided with disconnectors*

Interlocks shall be provided to prevent operation of disconnectors under conditions other than those for which they are intended (refer to IEC 62271-102). The operation of a disconnector shall be prevented unless the circuit-breaker, switch or contactor is in the open position.

Note 1 This rule may be disregarded if it is possible to have a busbar transfer in a double busbar system without current interruption.

The operation of the circuit-breaker, switch or contactor shall be prevented unless the associated disconnector is in the closed, open or earthing position (if provided).

The provision of additional or alternative interlocks shall be subject to agreement between manufacturer and user. The manufacturer shall give all necessary information on the character and function of interlocks.

Earthing switches having a rated short-circuit making capacity less than the rated peak withstand current of the main circuit should be interlocked with the associated disconnectors.

Apparatus installed in main circuits, the incorrect operation of which can cause damage or which are used for securing isolating distances during maintenance work, shall be provided with locking facilities (for example, provision for padlocks).

If earthing of a circuit is provided by the main switching device (circuit-breaker, switch or contactor) in series with an earthing switch, the earthing switch shall be interlocked with the main switching device. Provision shall be made for the main switching device to be secured against unintentional opening, for example, by disconnection of tripping circuits and blocking of the mechanical trip.

Note 2 Instead of an earthing switch, any other device in its earthing position is possible.

If non-mechanical interlocks are provided, the design shall be such that no improper situations can occur in case of lack of auxiliary supply. However, for emergency control, the manufacturer may provide additional means for manual operation without interlocking facilities. In such case, the manufacturer shall clearly identify this facility and define the procedures for operation.

7.2 Phase Color Code and Phase Sequence

China national standard GB 50060 and China electrical industry standard DL 5352 stipulate the phase color of each power source:

When facing the outgoing line, from left to right, from far to near, from top to bottom, the phase sequence is A, B, C. For the hard conductor inside the house and bus duct outside the house, there should be phase signs. A, B and C phase color code should be yellow, green and red respectively. The expansion works should be consistent with the original color code.

Q/GDW 11162 [31] specified the interface color of substation monitoring system. If there is no special requirement for the actual project, this standard can be used as a reference and details shall refer to Table 24.

DL 5136 specified the color of analog busbars and small busbars of the electrical system, and Q/GDW 11162 specifies the color code of the monitoring system, which are listed in this report for reference.

BS 7671 [32] defines the phase color of AC electrical system, DC low system and control system respectively. For example, the phase color of L1/2/3 ac three-phase system is brown, black and gray, which is greatly different from the typical domestic practice.

IEC 60445 [33] defines the phase color of ac system as L1/2/3 (conductor) & U/V/W (terminal) as black, brown and gray, which is inconsistent with the national

Table 24 Monitor system graphical interface color code

Rated voltage	Color	Red (R)	Green (g)	Blue (B)
1000 kV	Blue	0	0	225
750 kV	Orange	250	128	10
500 kV	Red	250	0	0
330 kV	Bright blue	30	144	255
220 kV	Purple	128	0	128
110 kV	Vermeil	240	65	85
66 kV	Gold	255	204	0
35 kV	Yellow	255	255	0
20 kV	Brown	226	172	6
10 kV	Light green	0	210	0
6 kV	Dark blue	0	0	139
0.4 kV	Tan	210	180	140
Background	Black	0	0	0
Equipment power off	Dark grey	128	128	128
Not put into operation and future bay	Light grey	192	192	192

standard of yellow, green and red, and consistent with BS 7671. During practice, color code shall be consient with local requirement and existing project (Tables 25, 26, 27, 28, 29, 30, and 31).

IEC 60445 "Basic and safety principles for man–machine interface, marking and identification-Identification of equipment terminals, conductor terminations and conductors".

7.3 Battery

7.3.1 Battery Cabinet and Battery Bank

As specified in DL 724 [35], when nickel–cadmium batteries are 40 Ah and below, and valve controlled batteries are 300 Ah and below, the batteries can be placed in the cabinet.

DL 5044 provides for the setting up of separate battery rooms as per followings:

- Valve controlled lead-acid battery 300 Ah or above;
- Fixed type exhaust lead-acid battery;
- 100 Ah or above medium rete nickel–cadmium alkaline batteries.

Table 25 Mimic panel bus color code

No.	Voltage (kV)	Color
1	DC	Brown
2	AC 0.10	Light grey
3	AC 0.23	Dark grey
4	AC (0.40)	Ochre yellow
5	AC 3	Dark green
6	AC 6	A deep brilliant blue
7	AC 10	Iron oxide red
8	AC 13.8	Light green
9	AC 15.75	Green
10	AC 18	Pink
11	AC 20	Iron yellow
12	AC 35	Lemon yellow
13	AC 63	Orange
14	AC 110	Scarlet
15	AC (154)	Brilliant blue
16	AC 220	Purple red
17	AC 330	White
18	AC 500	Light yellow

Notes
(1) This table is the color code specified in the current industry standard "General Technical Conditions of Control and Relay Protection Screen (Cabinet, Platform) for Secondary Circuits of Power System" JB/T5777.2 [34]
(2) Width of mimic bus should be 12 mm
(3) Color code for the excitation system DC mimic bus shall refer to No.1 color
(4) The color of the transformer neutral mimic bus is black
(5) AC 750 kV (1000 kV) temporarily use medium blue
(6) The voltage class in brackets is the non-standard voltage value

7.3.2   Battery Room Explosion-Proof Requirement

(1) *GB 50058* [36]:

(1)   When a closed area containing rechargeable nickel–cadmium or nickel-hydrogen batteries without vent, and battery total volume is less than 1% of the closed area, and the capacity of the battery is less than 1.5 Ah at the 1-h discharge rate, then the battery area can be considered as a non-hazardous area.

(2)   When a closed area containing other batteries other than above doesn't have, and battery total volume is less than 1% of the closed area, or the rated output of the charging system of the battery is less than or equal to 200 W and measures

Table 26 Small bus color code [4]

Symbol	Name	Color
+KM	Control small bus (positive power)	Red
−KM	Control small bus (negative power)	Blue
+XM	Signal busbar (positive power)	Red
−XM	Signal busbar (negative power)	Blue
(+)SM	Flash bus	Red, green with interval
YMa	Low voltage bus (A phase)	Yellow
YMb	Low voltage bus (B phase)	Green
YM	Low voltage bus (C phase)	Red
YMN	Low voltage bus (zero line)	Black

are taken to prevent inappropriate overcharging, etc., then the battery area can be considered as non-hazardous area.

(3) Non-enclosed areas containing rechargeable batteries with well ventilation, then the area can be classified as non-hazardous.

(4) When all batteries can directly or indirectly exhaust to the outside of the enclosed area, the area can be considered as a non-hazardous area.

(5) When a closed area equipped with a battery and poorly ventilated has a ventilation condition that at least can ensure that the ventilation condition of the area is not less than 25% that meets the good ventilation condition and the charging system of the battery has a design to prevent overcharging, it can be classified as zone 2. If the condition is not satisfied, it can be classified as zone 1.

(2) *API RP 505* [37]:

The principle of the hazardous area classification of battery room is as follows:

This section presents guidelines for classifying locations where batteries are installed. Areas classified solely because they contain batteries are classified because of hydrogen evolution from the batteries and therefore require a Group IIC design.

(1) Areas containing non-rechargeable batteries do not require area classification solely due to the presence of the batteries.

(2) Do not require area classification solely due to the presence of the batteries, when enclosed areas installed rechargeable batteries that:

(a) Without vents;

(b) Be of the nickel–cadmium or nickel-hydride type;

Table 27 Color code as per BS 7671 [32]

Function	Alphanumeric	Color
Protective conductors		Green-and-yellow
Functional earthing conductor		Cream
a.c. power circuit[1]		
Phase of single-phase circuit	L	Brown
Neutral of single- or three-phase circuit	N	Blue
Phase 1 of three-phase a.c. circuit	L1	Brown
Phase 2 of three-phase a.c. circuit	L2	Black
Phase 3 of three-phase a.c. circuit	L3	Grey
Two-wire unearthed d.c. power circuit		
Positive of two-wire circuit	L+	Brown
Negative of two-wire circuit	L−	Grey
Two-wire earthed d.c. power circuit		
Positive (of negative earthed) circuit	L+	Brown
Negative (of negative earthed) circuit[2]	M	Blue
Positive (of positive earthed) circuit[2]	M	Blue
Negative (of positive earthed) circuit	L−	Grey
Three-wire d.c. power circuit		
Outer positive of two-wire circuit derived from three-wire system	L+	Brown
Outer negative of two-wire circuit derived from three-wire system	L−	Grey
Positive of three-wire circuit	L+	Brown
Mid-wire of three-wire circuit[2][3]	M	Blue
Negative of three-wire circuit	L−	Grey
Control circuits, ELV and other applications		
Phase conductor	L	Brown, Black, Red, Orange, Yellow, Violet, Grey, White, Pink or Turquoise Blue
Neutral or mid-wire[4]	N or M	

Notes
(1) Power circuits include lighting circuits
(2) M identifies either the mid-wire of a three-wire d.c. circuit, or the earthed conductor of a two-wire earthed d.c. circuit
(3) Only the middle wire of three-wire circuits may be earthed
(4) An earthed PELV conductor is blue

Table 28 Color code as per IEC 60445

Designated conductors/terminals		Identification of conductors/terminals by			
		Alpha numeric notations[a]		Colours	Graphical symbol[b]
		Conductors	Terminals		
AC conductors		AC	AC	–	–
1	Line 1	L1	U	BK[d] or	
2	Line2	L2[c]	V	BR[d] or	
3	Line3	L3[c]	W	GR[d] or	
4	Mid-point conductor	M	M	BU[e]	No recommendation
5	Neutral conductor	N	N		
DC conductors		DC	DC	–	
1	Positlve	L+	+	RD	
2	Negative	L–	–	WH	
3	Mid-point conductor	M	M	BU[e]	No recommendation
4	Neutral conductor	N	N		
Protective conductor		PE	PE	GNYE	
1	PEN conductor	PEN	PEN	GNYE BU	No recommendation
2	PEL conductor	PEL	PEL		
3	PEM conductor	PEM	PEM		
Protective bonding conductor[g]		PB	PB	GNYE	
1	– Earthed	PBE	PBE		
2	– Unearthed	PBU	PBU		No recommendation
Functional earthing conductor		FE	FE	PK	
Functional bonding conductor		FB	FB	No recommendation	

    (c)   With a total volume less than one-hundredth of the free volume of the enclosed area; and

    (d)   With a capacity not exceeding 1.5 A-hours at a 1 h discharge rate.

(3)   Do not require area classification solely due to the presence of the batteries, when enclosed areas containing rechargeable batteries that:

Table 29 Reference value of leakage current for 220/380 V single-phase and three-phase line laying through conduit (mA/km)

Insulation material	Wire conductor section (mm^2)												
	4	6	10	16	25	35	50	70	95	120	150	185	240
XLPE	52	52	56	62	70	70	79	89	99	109	112	116	127
Rubber	27	32	39	40	45	49	49	55	55	60	60	60	61
PVC	17	20	25	26	29	33	33	33	33	38	38	38	39

      (a)   Without vents.

      (b)   have a total volume less than one-hundredth of the free volume of the enclosed area, or have a charging system that has a rated output of 200 watts or less and that is designed to prevent inadvertent overcharging do not require area classification solely due to the presence of the batteries.

(4)   A non-enclosed adequately ventilated location containing batteries is unclassified.

(5)   An enclosed location containing rechargeable batteries is unclassified provided all batteries are vented either directly or indirectly to the outside of the enclosed area.

(6)   An enclosed, adequately ventilated location (excluding battery boxes) containing batteries is classified as follows:

    – Unclassified provided (a) calculations verify that natural ventilation will prevent the accumulation in the enclosed location of hydrogen above 25% of its LFL during normal float charge operations, and (b) the battery charging system is designed to prevent inadvertent overcharging.

    – Unclassified provided (a) calculations verify that mechanical ventilation will prevent the accumulation in the enclosed location of hydrogen above 25% of its LFL during normal float charge operations, (b) the battery charging system is designed to prevent inadvertent overcharging, and (c) effective safeguards against ventilation failure are provided.

    Ventilation rates should be based on the maximum hydrogen evolution rate for the applicable batteries. Lacking specific data, the maximum hydrogen evolution rate for all batteries should be considered as $1.27 \times 10\text{–}7$ m^3/s (0.000269 ft^3/min) per charging ampere per cell at 25 °C, and standard pressure (101.325 kPa) with the maximum charging current available from the battery charger applied into a fully charged battery.

(7)   An enclosed, inadequately ventilated area containing batteries is classified as follows:

    – Zone 2, provided (a) ventilation is at least 25% that required for adequate ventilation, and (b) the battery charging system is designed to prevent inadvertent overcharging.

    – Zone 1, if the criteria specified above is not met.

Table 30 Reference value of leakage current for motor

Motor rated rating (kW)	1.5	2.2	5.5	7.5	11	15	18.5	22	30	37	45	55	75
Normal leakage current (mA)	0.15	0.18	0.29	0.38	0.50	0.57	0.65	0.72	0.87	1.00	1.09	1.22	1.48

Table 31 Reference value of leakage current for typical electrical appliances

Equipment	Leakage current (mA)
Computer	1–2
Printer	0.5–1
Small mobile electrical appliance	0.5–0.75
Teleprinter	0.5–1
Copier	0.5–1.5
Filter	1
Fluorescent lamps mounted on metal structure	0.1
Fluorescent lamps mounted on non-metallic structure	0.02

Note when calculate total leakage current for different equipment, correction factor (0.7/0.8) shall be applied

- Zone 0 classification would normally prohibit the installation of batteries in the area. Check applicable requirements.

7.4 Residual Current Protection

7.4.1 Poles Selection of Residual Protection Device

JGJ 16 specifies that when installing residual current protection device, it shall be able to disconnect all live conductors in the protected circuit. The N line is a live conductor so all residual current protection device shall disconnect N line. Numbers of residual current can be summarized as follows:

- Single-phase power feeder: 2P (L + N).
- Three-phase motor feeder: 3P (L1 + L2 + L3).
- Three-phase power feeder: 4P (L1 + L2 + L3 + N).

7.4.2 $I_{\Delta n}$ of Residual Protection Device

As per GB51348 [38], the following distribution circuits shall be protected by residual current protection device with rated residual current no more than 30 mA (civil building electrical design):

- Hand-held and mobile electrical equipment;
- Stationary devices that the human body may not be able to get rid of in time;
- Electrical equipment in outdoor workplaces;
- Household electrical appliance circuit or socket circuit.

As per Distribution Manual (Revision 4th) specifies the rated residual current for earth fault protection device shall be as followings:

- More than twice of the distribution feeder normal leakage current;
- The minimum non-action time of RCD on the power side should be greater than the total action time of RCD on the load side;
- The $I_{\Delta N}$ of the power side RCD should be at least 3 times than that of the load side RCD.

References

1. GB 50052-2009, Code for design electric power supply systems, Section 3.0.
2. GB 50350-2015, Code for design of oil- gas gathering and transportation systems of oil field, Section 11.1.
3. IEC 60364-1-2005, Low-voltage electrical installations – Part 1: Fundamental principles, assessment of general characteristics, definitions, Section 35.2.
4. DL 5136-2012, Technical code for the design of electrical secondary wiring in fossil-fired power plants and substations, Section 2.0.
5. Industrial and Civil Power Supply and Distribution Design Manual, fourth edition
6. Alan L. Sheldrake. Handbook of Electrical Engineering for Practitioners in the Oil, Gas and Petrochemical Industry [M]. 2002, Section 2.0.
7. IEEE 3002.3-2018, Recommended Practice for Conducting Short-Circuit Studies and Analysis of Industrial and Commercial Power Systems, Section 4.0.
8. DL 5222-2005, Design technical rule for selecting conductor and electrical equipment, Section 5.0.
9. DEP 33.64.10.10 Electrical Engineering Design. Section 3.6
10. GB 50049-2011, Code for design of small fossil fired power plant, Section 17.0.
11. API 540-2013, Electrical Installations in Petroleum Processing Plant, Section 4.3.
12. IEEE 3006, Historical Reliability Data for IEEE 3006 Standards: Power Systems Reliability. Part 1.
13. IEEE 3006.5-2014, Recommended Practice for the Use of Probability Methods for Conducting a Reliability Analysis of Industrial and Commercial Power Systems. Section 3.0~10.0.
14. GB/T 50064-2014, Code for design of overvoltage protection and insulation coordination for AC electrical installations, Section 4.4.
15. NB/T 35067-2015, Overvoltage protection and insulation coordination design guide for hydropower station, Section 4.0.
16. ABB 《High Voltage Surge Arresters Buyer's Guide》, Table 1.
17. ABB 《Dimensioning, testing and application of metal oxide surge arresters in medium voltage networks》, Section 4.0~6.0.
18. Substation Design Application Guide. Section 3.9.
19. DL/T 5408-2009, Specifications for disposition, erection and acceptance of 220/380V power surge protection in power plant and substation. Section 5.3.
20. GB 50057-2010, Code for design protection of structures against lightning. Section 6.4.
21. GB 50060-2008, Code for design of high voltage electrical installation (3 ~ 110kV), Section 5.0.
22. IEC 61936-1-2014, Power installations exceeding 1 kV a.c. – Part 1: Common rules, Section 5.4.
23. DL/T 5352-2018, Code for design of high voltage electrical switchgear. Section 5.4.
24. GB 50054-2011, Code for design of low voltage electrical installations. Section 4.2.
25. DL/T 5153-2014, Technical code for the design of auxiliary power system of fossil-fired power plant. Section 7.2.

26. NFPA 70-2020, National Electrical Code. Section 110.31.
27. IEEE C2-2017, National Electrical Safety Code. Rule 180B.
28. GB 50229-2019, Standard for design of fire protection for fossil fuel power plants and substations. Section 11.1.
29. IEEE 979-2012, Guide for Substation Fire Protection. Section 7.2.
30. IEC 62271-200-2011, High-voltage switchgear and controlgear – Part 200: AC metal-enclosed switchgear and controlgear for rated voltages above 1 kV and up to and including 52 kV. Section 5.11.
31. Q/GDW 11162-2014, Graphic interface specification for supervision and control system of substation. Annex A.
32. BS 7671-2018, Requirements for Electrical Installations IET Wiring Regulations Eighteenth Edition. Appendix 7.
33. IEC 60445-2010, Basic and safety principles for man-machine interface, marking and identification – Identification of equipment terminals, conductor terminations and conductors. Annex A.
34. JB/T 5777.2-2002General specification for control and protection panel (cabinet desk) of secondary circuit of power system. Section 5.3.
35. DL/T 724–2000, Specification of operation and maintenance of battery DC power supply equipment for electric power system. Section 4.4.
36. GB 50058-2014, Code for design of electrical installations in explosive atmospheres. Annex B.
37. API 505-2018, Recommended Practice for Classification of Locations for electrical Installations at Petroleum Facilities Classified as Class I, Zone 0, Zone 1, and Zone 2. Section 8.2.6.
38. GB51348-2019, Standard for electrical design of civil buildings. Section 7.5.

Chapter 3
Safety Analysis for Overhead Transmission Line Design

With the development of international oil field surface engineering, the power demand of each substation and switchyard is daily increased. The production department has more and more serious technical and commercial requirements on power supply, which not only need to meet the sufficient power demand, but also to guarantee its safety and reliability for the transmission lines. As the focal point of oil field power system, transmission line design plays the important role in the operation stability of the power grid system. The intrinsic design safety, environmental meteorological condition and man-made damage all have the significant impacts on the reliability of the system.

In this chapter, the safety of transmission line design is introduced from the safety coefficient selection, lightning protection and safety protection of transmission line and other points related to intrinsic safety.

1 Safety Coefficient for OHTL

1.1 Safety Coefficient Selection

The safety coefficient of overhead transmission lines is used to reflect the safety level of lines and frames in engineering structure design. The determination of safety coefficient needs to consider various uncertainties such as load, mechanical material stress, difference between test value and design value and actual value, calculation mode and construction quality.

The safety coefficient relates to the economic benefits of the project and the possible consequences of structural damage, such as life, property, and social influence. In the design process of overhead transmission lines, the safety coefficient mainly involves important aspects as the operating tension of the earthing conductor, the strength of the fittings and insulators, the tower structure, and the stress on the

© Petroleum Industry Press 2022

K. Ma et al., *International Oilfield Surface Facilities: Safety Analysis for Electrical Design*, https://doi.org/10.1007/978-981-16-3104-7_3

foundation, etc., which are important parameters for the line design, operation and maintenance safety.

1.1.1 Safety Coefficient of Fittings and Insulators

After the type of insulator string is determined, it is necessary to select the insulator and the fittings to meet the mechanical damage load of the component fittings. Mechanical failure load refers to the maximum load that can be achieved during the FAT conditions (laboratory test conditions: frequency voltage shall be processed on insulator string and the voltage shall be maintained during the test, at the same time tensile load shall be applied on insulator string).

The metal materials used for insulators and fittings are like those used for the mechanical fittings. The design method used for the mechanical unit is the safety coefficient design methodology; similarly the safety coefficient design procedure should also be used in the design of insulators and fittings.

Many kinds of insulators are adopted for towers and poles. Different insulators have different operating environments and conditions, and their safety coefficients are also different. In the selection of OHTL fittings, appropriate insulators and fittings should be selected according to the line voltage level and its importance in the system. Suitable insulators and fittings can not only ensure their own safety, but also guarantee the safety of the whole overhead transmission line.

Pin insulators and porcelain crossarm insulators are usually used for straight line pole with voltage up to 10 kV with easy maintenance and lower cost.

Butterfly insulator is with higher cost in comparison of pin insulator but with more safety level, is commonly used for tension and straight tower/pole with voltage up to 10 kV and above.

Suspension insulator strings shall be adopted on lines of 33 kV and above. In case of a single insulator damage, power system can keep operation until the broken point is found out and to save time for maintenance, also to reduce tension on tower through the deflection of insulator strings in case of single wire broken, thereby ensuring the safety and stability of the tower. When passing through high polluted areas, anti-pollution insulators and composite insulators should also be considered for enhancing insulation level for particular area.

The mechanical strength design of insulators and fittings is shown in formula (1)

$$KF < F_n \tag{1}$$

Note:

K—Safety coefficient of mechanical strength

F—Design load, kN;

F_n—Mechanical failure load of suspension insulators, butterfly insulators and fittings; bending limit of pin insulators and porcelain crossarm insulators, kN (Table 1).

Table 1 National standards on the safety coefficient of insulators and fittings [1]

Insulator type	Safety coefficient	
	Normal	Broken wire
Suspension insulator	2.7	1.8
Pin insulator	2.5	1.5
Butterfly insulator	2.5	1.5
Porcelain crossarm insulator	3	2
Fittings	2.5	1.5

In consideration of different environmental condition, government regulation, geographical condition, the variance of safety coefficient for transmission line insulator fitting, foundation design shall be analyzed seriously, to avoid further serial impact on operation and maintenance of OHTL. The duration, cost and difficulties shall be too high to replace insulator after completion of construction. Designers should fully understand the local environmental conditions and to consult the design input documents to find if there is any local reference on the safety coefficient of insulators and fittings. It is strictly prohibited to apply domestic standards to any international projects directly other than the same working condition (Table 2).

In common engineering practice in recent years, the safety coefficient of insulators and fittings in Invitation to Bid documents (ITB) is specified and to be complied.

Table 2 Safety coefficients for OHTL of different countries

Country	Standard	Strength design mode	Safety coefficient (maximum load permitted)		Note
			Insulator	Fitting	
USA	NESC(1977)	A	2.0~2.5	–	By load type
	B.P.A	B	(100%RUS)	–	
Canada	CSA-C223(1970)	A	2.0	–	
	Ontario Hydro	B	(60% 85%RUS)	(60% 85%RUS)	By load note
	Hydro Quebec	B	(70%RUS)	–	
France	Standard in (1970)	A	3.0	–	
	EDF	B	(60%RUS)	(60%RUS)	Ice
Germany	VDE0210(1969)	A	3.0~3.6	2.5~5.0	By fitting and insulator type
Sweden	SEN-3601(1961)	A	2.0~3.0	2.0	
Russia	(1985)	A	2.7	2.5	
Japan	JEAC6001(1978)	A	2.5	2.5	

Oil fields have a wide geographic distribution area, also with different natural environment and operational conditions, insulator selection must be considered in combination with different operational environment requirements, example in central Asia cold areas composite insulator should not be adopted, porcelain insulator and composite insulator should not be selected for high temperature area of Africa, toughen glass insulator is mostly recommended in these hot temperature area.

1.1.2 Safety Coefficient of Conductor as GB50061 [2]

The conductor is the most important part of the overhead transmission line design. The final goal of all the OHTL design is for the conductor to be able to safely deliver power to end user. The safety of the conductor is divided into two aspects: one hand, the conductor have to ensure its own safety, to ensure its own strength and current capacity must meet the requirements, on the other hand, the conductor have to ensure the safety of human beings and obstacles to meet the safety distance.

Safety coefficient of conductor is refer to the ratio between the breaking load and the maximum tension strength of the wire at the lowest point of sag, or the ratio between the breaking stress and the maximum tension stress at the lowest point of sag. The selection of conductor safety coefficient is directly corresponding to the safety and economic operation of transmission line. The design code stipulates that the safety coefficient of the conductor shall not be less than 2.5.

In extreme weather conditions with rare wind speed or rare ice-covered, the maximum horizontal conductor tension load shall not exceed 60% of the breaking strength. The maximum tension force of the hanging point shall not exceed 66% of the breaking strength.

The shield wire mostly adopts steel wire, it is easy to be corroded, and its design safety coefficient should be higher than that of conductor. OPGW (Optical Fiber Composite Overhead Ground Wire) is adopted in most projects of international oil fields, and its composition is similar with ACSR (aluminum conductor steel reinforced). Therefore, when selecting the safety coefficient, it should be matched from sag, tension and other aspects to select the appropriate coefficient. The safety coefficient of OPGW should be higher than or equal to the safety coefficient of the conductor (Table 3).

1.1.3 Safety Coefficient in Wind Load Calculation

Wind load is the main external load on conductor. The standard value of wind load on conductor and shield wire under vertical wind should be calculated according to Formula (2)

$$W_x = \alpha \ U_s dL_W W_o \qquad (2)$$

Note: W_x—Standard value of wind load on conductor, kN;

Table 3 Safety coefficient of conductor & shieldwire on stright support

Type of conductor		Breaking strength to natural tension %		
		Concrete poles and steel pole	Tower with stay wire	Free standing tower
Shield wire		15 - 20	30	50
Conductor	Less than 95 mm^2	30	30	40
	Between 120 ~ 185 mm^2	35	35	40
	More than 210 mm^2	40	40	50

Table 4 Wind speed conversion coefficient

Wind speed interval	1 h	10 min	5 min	2 min	1 min	30 s	20 s	10 s	5 s	3 s	Instant
Conversion coefficient	0.94	1	1.07	1.16	1.20	1.26	1.28	1.35	1.39	1.42	1.50

α—Wind load span coefficient.

d—sum of outer diameters of conductor or shield wire with icing (for split wires, Interline effects on shielding should not be considered), m;

U_s—coefficient of wind load 1.2 with D < 17 mm, 1.1 with D ≥ 17 mm, and 1.2 with ice.

L_w—Wind span;

W_o—standard value of wind pressure;

The wind during a 10 min period at a level of 10 m above ground shall be used for calculation, but as the reference data available from meteorological departments around the world are not same, it need to convert for calculation in most of cases. According to the load code of building structure, the conversion coefficient of each wind speed interval is shown in Table 4.

1.1.4 Safety Coefficient of Tower Structure Design as GB50061 [3]

(1) Tower structure calculation in domestic national standard

In domestic national standard, the ultimate state design formula is expressed by load standard value, material performance standard value, geometric parameter standard value and various sub-item coefficient.

As foumula (3):

$$\gamma_0(\gamma_G \times S_{GK} + \psi \Sigma \gamma_{Qi} \times S_{QiK}) \leq R \tag{3}$$

γ_o—The importance coefficient of tower structure and the important lines should not be less than 1.1, 0.9 for temporary lines and 1.0 for other lines.

γ_G—The partial coefficient of permanent load should not be greater than 1.0 when it is favorable to the structure, and 1.2 when it is unfavorable.

γ_{Qi}—The component coefficient of variable load of item i, is 1.4;

S_{GK}—Effect of permanent load standard value;

S_{QiK}—The effect of variable load standard value of item i;

ψ—Variable load combination coefficient, 1.0 for normal operation, 0.9 for accident, installation, and uneven ice, and 0.75 for checking;

R—Resistance design value of structural members.

(2) American standard tower structure calculation

See formula (4):

$$\varphi R_n > \text{effect of [DL and } \gamma Q_{50}] \qquad (4)$$

f—Material strength coefficient (determined by component reliability factor LEL, f = 1.0 when LEL = 0.1, F = 0.95 when LEL = 2, F = 0.90 when LEL = 20, f = 0.86 when LEL = 50);

R_n—Nominal yield strength of a member

DL—The permanent load

γ—The loading coefficient applied to the loading effect Q50

Q_{50}—Design load value for 50 year recurrence period.

By comparison, the American standard stipulates that the design strength of the member shall be the smaller of the two values obtained in the two limit states of the gross section yield and the net section tension. However, it is based on the yield of the net section in China's national standards, which tends to be more reliable with higher cost.

1.1.5 Foundation Type and Its Safety Coefficient

The foundation of transmission lines is mainly to ensure the stability of the tower/pole. No matter in which extreme conditions, the tower shall be guaranteed stand still base on the stable support for the transmission line.

(1) Design of foundation type

The foundation type of transmission line should be determined according to the type of poles and towers, topography along the line, engineering geology, hydrology, as well as construction and transportation conditions, etc. The detail design of foundation should be mainly depending on civil engineering, with the assistance of electrical.

The basic foundation classification is as follows:

(1) Large excavation foundation;

(2) Excavate and expand the bottom foundation;
(3) Blasting and expanding pile foundation;
(4) Rock anchor pile foundation;
(5) Bored pile foundation;
(6) Overturn the foundation.

Large excavation foundation, excavating and expanding foundation and bored pile foundation of overhead transmission line are most adopted in oil field OHTL engineering.

(2) Safety coefficient in foundation design

There are two kinds of safety coefficients in foundation design: one is safety coefficient in up-lifting and overturning design, and the other is safety coefficient in strength design. In the process of project design, the electrical engineering should provide the loads to the civil engineers, and the civil engineers should select the relevant safety coefficient to ensure the effectiveness of the coefficient and the economy of the foundation. The specific safety coefficients are shown in Table 5 and 6.

Table 5 Up-lifting and overturning design safety coefficient as GB50061 [4]

Tower type	UP reliability		Overturning reliability
	Soil weight	Foundation weight	
Suspension	1.6	1.2	1.5
Tension	2.0	1.3	1.8
Terminal or long span	2.5	1.5	2.2

Table 6 Strength design safety coefficient

	Mechanical characteristics	Safety coefficient
Concrete	Compression member calculated according to its compressive strength, local pressure	1.7
	Member under compression, as calculated by compressive strength	2.7
Reinforced concrete	Eccentric tension (pressure), bending, torsion, local pressure,	1.7
	Percussion cut, none	2.2

Fig. 1 Clearance circle
check for tower/pole

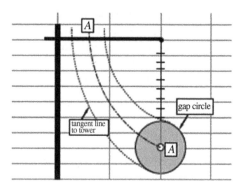

1.2   Safety Clearance of Transmission Line

1.2.1   Safety Clearance and Clearance Circle Check

The main function of insulated coordination for transmission line is to make sure
that the line to operate safely and reliably under various conditions such as power
frequency voltage, operating overvoltage, and lightning overvoltage and so on. The
clearance circle drawing based on the safety clearance between live part and dead
part is a common inspection method. Regarding the clearance circle, it takes the
conductor axis as the center and the radius of the conductor, safety clearance and
margin as the radius, put the tower head elements and fitting deflection angle into
a simple clearance circle diagram. If tower elements are outside the tangent line to
tower, it means the clearance of the tower can meet the insulation requirements, as
shown in Fig. 1.

The clearance circle is simple, but it is an indispensable check method in the
OHTL design. It can make the clearance between the conductor and the dead part
concise and explicit. If maintenance under live condition is required, the range of
activities of worker should be added to the clearance circle. Whatever single circuit
tower, double circuit tower or complex four-circuit tower, all should be checked by
clearance safety circle.

1.2.2   Safe Clearance Between Conductors and Obstacles as GB50061
[5]

(1) Vertical safety clearance and horizontal safety clearance

The safety clearance between conductor and obstacle includes vertical safety
clearance and horizontal safety clearance.

The vertical safety clearance determines the minimum/maximum height of the
conductor and plays a critical role in the design of the pole/tower. Horizontal safe
clearance played on important role on OHTL route. See Table 7, 8, 9, 10 and 11 for

Table 7 Minimum clearance from conductor to ground

Area	Minimum distance		
	Voltage level		
	3 kV and below	3~10 kV	35~66 kV
Densely populated region	6.0	6.5	7.0
Sparsely populated region	5.0	5.5	6.0
Limited transportation region	4.0	4.5	5.0

Table 8 Minimum clearance from conductor to slopes, cliffs, or rocks

Area	Minimum distance		
	Voltage level		
	3 kV and below	3~10 kV	35~66 kV
Slopes can be reached	3.0	4.5	5.0
Slopes, cliffs, or rocks cannot be reached	1.0	1.5	3.0

Table 9 Minimum clearance from conductor to slopes, cliffs, or rocks

Voltage level	3 kV and below	3~10 kV	35 kV	66 kV
Distance	3.0	3.0	4.0	5.0

Table 10 Minimum clearance from outer conductor to building structure

Voltage level	3 kV and below	3~10 kV	35 kV	66 kV
Distance	1.0	1.5	3.0	4.0

the safe clearance between various obstacles and conductor.

(2) Clearance for live maintenance

If live maintenance is required for OHTL, the maintenance distance should be considered in tower design, as shown in Table 12.

(3) Minimum clearance between phase conductors

The clearance between wires shall be determined according to the following requirements and combined with operating experience. The clearance of the horizontal span below 1000 m shall be calculated according to Formula (5):

Table 11 Minimum clearance from conductor stress

Voltage level	3 kV and below	3 ~ 10 kV	35 ~ 66 kV
Distance	3.0	3.0	4.0

Table 12 Minimum clearance from live working part to tower earthing part

Voltage level kV	Safety clearance m	Note
≤10	0.7	Clearance shall be expanded based on the actual condition
20,35	1.0	
66,110	1.5	
220	3.0	

$$D = 0.4L_kU + f_C \tag{5}$$

In formula:

D—Horizontal clearance between conductors, m;

L_k—Length of suspension insulator string, m;

U—Nominal voltage of transmission line, kV;

f_C—Maximum sag of conductor, m;

The distance between vertical conductors should be 75% of the calculated result in Formulas (3–5). For towers with suspension insulator strings, the clearance between vertical conductors should not be less than the values listed in Table 13.

The horizontal offset between the upper and lower layers of adjacent conductors or between the shield wire and adjacent conductors in the ice-covered area (if there is no operating experience) should not be less than the values listed in Table 14.

The horizontal offset between the upper-layer and lower-layer adjacent conductors or between the shield wire and adjacent conductor can be reduced appropriately according to the operation experience in an area of 5 mm ice thickness. In heavy ice areas, the conductors should be arranged horizontally, and the horizontal offset value between the shield wire and the adjacent conductor should be increased by at least 0.5 m than the value of 15 mm designed ice thickness in Table 14.

Table 13 Minimum clearance between vertical conductors of straight towers

Rate voltage, kV	110	220	330	500
Vertical clearance, m	3.5	5.5	7.5	10.0

Table 14 Horizontal offset between adjacent conductors on the upper and lower layers or between ground wires and adjacent conductors

Rating voltage, kV		110	220	330	500
Horizontal offset, m	10 mm, ice thickness	0.5	1.0	1.5	1.75
	15 mm, ice thickness	0.7	1.5	2.0	2.5

2 Lightning Protection for Overhead Transmission Lines

The operation experience of international s oil field shows that lightning strike is main reason for trip of overhead transmission line. The probability of OHTL tripped by lightning strike is closely related to the local meteorological conditions, designers must investigate the design concept of existing reliable lines in the local area and investigate the thunderstorm days and levels in different seasons in order to avoid the impact of lightning trip on the overall power supply reliability of OHTL.

2.1 Lightning Strike and Thunderstorm Day

2.1.1 Direct Lightning

In the main discharge stage of the lightning to tower, the negative charge in the pilot channel neutralizes with the positive charge in the tower, shield wire and earth, forming the direct lightning impulse current. On the one hand, the negative lightning shock wave leads down through the tower and to both sides along the shield wire, making the potential on the top of the tower rise and changing the conductor potential through electromagnetic coupling. On the other hand, the rapid development of positive polar lightning waves from the top of the tower to the thunder cloud causes the rapid change of the space electromagnetic field, and the positive polar induced lightning waves appear on the conductors. The voltage acting on the line insulator string is the potential difference between the tower top and the conductor. Once this voltage exceeds the impulse discharge voltage of the insulator string, the insulator string flashover occurs. Here, the tower potential is higher than that of the conductor; hence it is often called "counterstrike."

2.1.2 Induced Lightning

Induced thunder, also known as lightning induction, when the thunder cloud is close to the overhead line, due to electrostatic induction, an equal number of opposite bound charge gathered, when the thunder cloud discharge to other places, the bound charge on the line is released and travel to both ends of the line, forming a high overvoltage. This high voltage will enter the room along overhead lines and metal pipes. According to the investigation and statistics, the lightning accidents caused by inductive lightning in the power supply system are account for 50~ 0% of the total lightning accidents.

2.1.3 Thunderstorm Days

Lightning protection design and lightning protection measures should be considered in oil field. The frequency of lightning activity in an area may be expressed as a thunderstorm days or thunderstorm hours in that area.

Local thunderstorm days or hours are likely to vary considerably from year to year, so multi-year average value should be investigated. In particular areas, such as some parts of the Middle East, thunderstorm days are not high, but are concentrated in January and February. Engineering design should pay special attention to such events that are short in duration, few in number, but high in frequency.

To sum up, the area with an average thunderstorm day of no more than 15 days a year is generally called a less thunderstorm area in China, which may not be completely applicable to international oil field surface projects. Lightning protection design should be tailored base on local conditions and treated differently.

2.2 Lightning Protection Design of Transmission Lines

2.2.1 Shield Wire

Transmission line lightning protection design should be based on the type of load and system operation condition in combination with the experience of local existing OHTL, intensity of regional lightning activity, topography and soil resistivity, etc., after calculating the lightning resisting level, and with technical and economic comparison to find the reasonable method of lightning protection.

The traditional lightning protection design of transmission lines includes shield wire, lightning arrester, coupling ground wire, amplification line insulation level, etc. The lightning arrester method adopted in Chinese project has achieved remarkable results in avoiding lightning strikes, but it is not recognized by the PMC and client of other countries in the design of international oil field. The common international practice is recommended to adopt shield wire.

The shield wire can lead the lightning current to the earth through shield wire, tower, and grounding device, to protect the object (conductor) from lightning strike. At the same time, it can also shunt the current into the ground and reduce the potential of the tower top. It has coupling effect on the wire to reduce the voltage on the insulator string when lightning strikes the tower. It has shielding effect to the conductor, may reduce the induction voltage on the wire.

The following protection methods shall be adopted for transmission lines in internationals oil fields:

(1) With reference to 132 kV and above lines, the probability of lightning strike is higher due to the high tower, so the shield wire should be set up along the whole line. During tower structure design, it is necessary to consider the protection angle of shield wire. Additionally, where the line located in high thunderstorm

Table 15 Power frequency earthing resistance of tower with shield wires as GB50061 [6]

Earthing resistivity, $\Omega \cdot m$	100 and below	>100~500	>500~1000	>1000~2000	>2000
Power–frequency grounding resistance, Ω	10	15	20	25	301

Note if soil resistivity more than 2000 Ω m, when grounding resistance is hard to reach 30 Ω, 6 ~ 8nos. radiate or continuous grounding devices which total length less than 500 m should be adopted, the grounding resistance is not restricted

day area and with higher importance grade, negative protection angle shall be adopted according to local regulations.

(2) The transmission lines of 66 kV and below shall be provided with a single shield wire along the whole line. In the areas where the average annual thunderstorm days are less than 15 days or the operation experience shows that lightning activity is slight, shield wire may not be set up for whole line. For power transmission lines without shield wire, it is advisable to set up 1–2 km shield wire in the inlet section of substation or power plant. The lightning protection angel is not more than 30°.

Except shield wire, there are other lightning protection methods, such as:

(1) Install automatic reclosing device
(2) Locally lightning arrester installation
(3) Improve the insulation level of the line and reduce the probability of flashover
(4) Reduce the earthing resistance of the tower, improve the level of lightning resistance, and reduce the probability of counterattack.

2.2.2 Earthing Resistance

In moist areas or where soil resistivity less than or equal to 100 Ω m, if natural earthing resistance of tower/pole is not more than Table 15, the natural grounding device (including stell tower foundation, and underground parts of reinforced concrete pole, such as dived pole and chassis, anchor plate, etc.) could be used directly without other artificial earthing device, except the terminal of power substation.

Where Soil resistivity between 100~300 Ω m, in addition to use natural ground of reinforced concrete pole/tower, artificial grounding device should be set up also. The buried depth of grounding body should not be less than 0.6~0.8 m.

In Soil resistivity between 300~2000 Ω m area, horizontal grounding device should be adopted, and the buried depth should not be less than 0.5 m. The earthing device in the cultivated land shall be buried below the working depth. Grounding devices in civilian areas and paddy fields, including temporary grounding devices, should be laid around the tower foundation to form a closed ring.

In areas where soil resistivity more than 2000 Ω m, 6–8 nos. radiate earthing devices in 500 m total length or continuous extension grounding system should be adopted. The buried depth of the earthing device should not be less than 0.3 m. The continuous extension shield wire is one or two ground wires embedded in the ground along the route and can be connected to the tower ground device of the next tower.

In areas with high soil resistivity, when radioactive grounding devices are used, if areas with low soil resistivity near the tower (within 1.5 times of the maximum length of each radioactive grounding device), external grounding or other measures may be adopted partly.

The cross-section and its shape of the grounding device have little influence on the earthing resistance value. Therefore, the selection of material specifications of the grounding device mainly considers corrosion and mechanical strength. Steel material usually is adopted in china and most of abroad projects chose copper.

Poles and towers with shield wires should be earthed. During the dry season when the thunderstorm days are concentrated, the power frequency grounding resistance of each tower without shield wire should not be higher than the values listed in Table 15. In areas with low soil resistivity, if the natural grounding resistance of the tower is not higher than the values listed in Table 15, extra grounding may not be installed.

International oilfield in the remote desert area, the soil resistivity is above 2000 Ω.m, the following measures should be considered during design.

(1) Resistance Reducing Agent

After laying resistance reducing agent around the ground electrode, it can increase the external size of the ground electrode, reduce the contact resistance between the ground electrode and the surrounding earth medium, and reduce the earthing resistance of the ground electrode to a certain extent, its resistance reduction effect is obviously when the resistance reducing agent is adopted for small area concentrated grounding and small grounding grid.

Resistance reducing agent is composed of several kinds of material preparation chemistry reducing agents, have good conductivity of electrolyte and water, these strong electrolyte and water surrounded by mesh colloid, spaces in mesh colloid filled by hydrolyzed colloid, this make it not run off with the groundwater and rain, and keep good conductive long-term effects, it is one of the latest and active approaches being adopted for grounding.

(2) Blasting grounding technology

Blasting grounding technology is a new technology developed in recent years to reduce the grounding resistance of grounding devices. It is used to make cracks by blasting and press machine to press low resistivity materials into the blasting cracks, to improve the conductivity of a large range of soil, which is equivalent to a large range of soil improvement.

(3) Horizontal grounding device with elongation

Combined with the practical application of engineering, the analysis results show that: when the length of the horizontal grounding body increases, the influence of inductance will increase, so as to increase the impact coefficient; when the length of the grounding body reaches a certain length, the impact grounding resistance will

Table 16 Soil resistivity corresponds to the length of horizontal grounding device

Soil resistivity, Ω·m	500	1000	2000
The effective length of the horizontal earthing, m	45~55	45~55	60~80

not drop. The effective length of the grounding body is determined according to the soil resistivity, as shown in Table 16.

Several other methods of reducing the earthing resistance of the tower are not commonly adopted in engineering, those methods are not particularly mentioned hereby, such as buried ground pole, sewage introduced into the ground device, deep well grounding, replacement of soil, chemical treatment of soil, etc.

Before determining the engineering method to reduce the resistivity of certain area, an integrated reasonable analysis shall be conducted based on original operating experience, characteristics of climate, topography, soil resistivity together with technical and economic conditions that it should not only keep its reliability and operability but also to guarantee the normal operation of the line, equipment, as the same time to avoid the too high grounding device engineering investment.

2.3 *Protection of Grounding System*

Due to the long distance of transmission line, Regular maintenance should be carried out during the service life after the grounding system is installed, which to avoid corrosion and theft.

(1) Corrosion

Copper corrosion rate is much slower than galvanized steel due to different metal activity. Therefore, in the selection of grounding materials, copper grounding devices should be adopted in the areas with severe corrosion to prevent rapid corrosion and ground failure.

The corrosion rate of the line grounding device is directly related to the corrosive rate of soil, the design of the line grounding system should be based on the soil resistivity and corrosion report in the geological survey report. For highly corrosive soil, if galvanized flat steel grounding device must be used, the grounding resistance should be tested annually and the whole line should be re-laid as soon as possible in case of a significant increase in the resistance value of even part of the line.

(2) Anti-theft

In some regions with relatively lower economic development, there are still serious cases of theft of power facilities. Since the shield wire bare and most easy to be found, also naturally become the most likely to lose.

According to the Statistics of the special project, the probability of shield wire theft of overhead lines outside the station is close to 100%, and there is no decreasing trend in recent years. Most project are located in highly corroded area, so copper ground wires had to be adopted to reduce the corrosion rate, and the locals living difficulties increased the probability of damage and theft of power facilities. Once the shield wire lost, there will be no effective ground of the tower and result in heavy impact on the power system or the surrounding personnel.

Simple and effective anti-theft measures are essential for the overhead lines out of the station. The main methods are as follows:

(1) Increases the difficulty of theft and reduces the profit of selling stolen goods; the exposed part does not use pure copper grounding wire to lower the price.
(2) Increase protection facilities for the ground wire, the shield wire directly lead out from the side of the foundation through the tower with steel conduit protection, or after installation, concrete to seal the shield wire directly on the foundation.

Different areas can adopt different measures based on the local situation, the effective way should be adopted to prevent grounding material lost, to ensure the effectiveness of the grounding system.

3 Auxiliary Safety Measures for Transmission Line

With exception of the above essential safety measures for OHTL itself, auxiliary safety measures are also necessary in the design of transmission lines, including ladder climbing, anti-climbing net, warning signs and bird protection measures.

3.1 Ladder

In order to ensure the convenience of maintenance, climbing ladders or foot nails should be set on the steel tower and steel pole. However, the setting of climbing ladder and foot nails also creates convenience for irrelevant personnel to climb the tower. Therefore, it should be noted that the climbing device should be set at a height that can avoid the danger of irrelevant personnel.

Domestic national standards stipulate that climbing ladders should be set above 2 m of the tower in consideration of domestic security condition, but may not be applicable for the requirements of anti-climbing in local oil field of other country, eg., according to the local regulations on overhead transmission line in special area, the tower should be equipped with climbing ladders higher than 5 m from the ground level, but it does make maintenance and repair much more difficult.

Safety is the primary factor for overhead transmission line. In consideration of experience in several projects and comprehensive consideration of various factor, it is advised that in transmission line projects of oil field, the ladder should be higher than 5 m away from the ground. Maintenance personnel can use portable overhaul ladder or lift truck for maintenance.

For lines requiring live maintenance, bilateral climbing ladders should be set to avoid the possibility of personnel entering the electrified side when climbing the tower.

3.2 Anti-climb Net

If irrelevant personnel climb to the tower or pole, there will be a high risk both for the line and for the climbers, which can lead to power failure even human being death. Therefore, in order to avoid irrelevant personnel climbing the tower and steel pole, anti-climbing devices should be set up for the tower.

There are many kinds of anti-climbing devices. The simplest way is to set up an anti-climbing net 3 m from the ground, barbed wire on the anti-climbing net can avoid irrelevant personnel approaching higher component. This kind of anti-climbing method is simple with low in cost, but the problem is that it may not play a good protective role against deliberate climbing behavior, and may even be destroyed by people, resulting in the failure of anti-climbing function.

There are various kinds of tower anti-climbing devices. Figure 2 is a kind of anti-climbing device that is often used in oil filed. For different projects, appropriate anti-crawling solutions should be selected to meet the multiple requirements of safety and cost.

Fig. 2 Anti-climbing device proposal

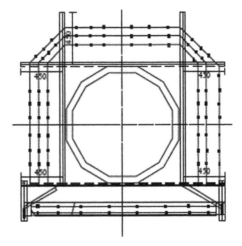

3.3 Warning Signs

Including anti-climbing facilities, warning signs should be set up to minimize the probability of accidents. Warning signs are mainly to prevent unrelated people getting too close to the line, or climbing the tower or pole. In potential danger areas, warning signs such as "high voltage" and "No climbing" must be installed, so that local people can clearly understand the dangers of the high voltage lines and stay away from transmission lines. If the client does not have specific requirements, the warning signs in Fig. 3 can be chosen for international oil field surface engineering design projects.

Fig. 3 Figure warning signs

3.4 Anti-bird Device

After years of operation experience of power transmission lines in oil field surface projects, anti-bird protection is a headache that must be considered for overhead transmission lines, and different regions face different bird potential danger to the operation reliability.

In the savanna of North Africa, variety of birds can be found everywhere all of the year. There is a kind of large bird, commonly known as garbage bird, with a wingspan of 2 m, which likes to stand on the crossarm and may cause short-circuit accident when flying to the air, also there are many small birds in the oil field area that like to nest in the corners of the towers, these bird is part of ecological balance should be protected but in other hand, the power supplying continuity also should be considered, the appropriate ecological anti-bird devices should be selected.

In the oil field of southwestern Asia, wheat and other crops are the best food for birds and its quantities are grown in large numbers. Every summer, thousands of groups small and medium-sized birds will come to feed in these areas, and a large number of birds would defecate on the upper part of the insulator, resulting in short circuit of the insulator string and other faults.

For different regions, designers need to adopt different anti-bird devices. Several facilities in Table 17 should be selected individually or parallel based on project requirements to minimize bird caused accident as much as possible.

Where bird damage is especially serious in particular oil field, it is difficult for traditional anti-bird measures to meet the requirements of birds disasters prevention. Increasing the safety clearance of wires will increase the line cost, but it can ensure the safety of overhead transmission line. Where safety clearance increasing is needed

Table 17 Corresponding table of bird damage and corresponding measures

Code	Accidents type caused by birds	Measures
1	Large birds stand between the upper wire and the cross arm	Install anti-bird thorn on the cross arms to prevent birds from standing on the cross arms
2	Small birds' nest in the corner of the tower	The dead corner area on the tower should be reduced as far as possible, and the wire in the dead corner area should be protected by heat shrinkable insulation belt. For very serious areas, artificial bird's nest can be considered as per induction
3	Birds defecate on insulators	Add anti-bird device to keep birds away from the insulator suspension points; Increase the creepage distance of the insulator; Increases Pollutant-resistance
4	A large gathering of birds	The use of electronic sound bird exorcist, effectively drive large Numbers of birds, prevent birds from gathering

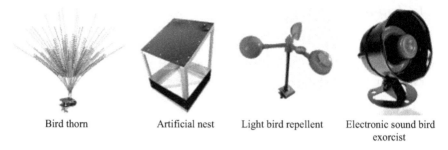

Bird thorn   Artificial nest   Light bird repellent   Electronic sound bird
                                                                            exorcist

Fig. 4 Figure warning signs

for reducing Birds disasters, design workshops should be held to determine the final solution (Fig. 4).

References

1. GB50061 Section 5.3- Insulator and fittings, Code for design of overhead power lines up to 66kV
2. GB50061 Section 8.1- Loads, Code for design of overhead power lines up to 66kV
3. GB50061 Section 9- Basic rules for tower structural design, Code for design of overhead power lines up to 66kV
4. GB50061 Section 11- Foundation Design, Code for design of overhead power lines up to 66kV
5. GB50061 Section 12- Locating, clearance to ground and crossing distance of overhead power line, Code for design of overhead power lines up to 66kV
6. GB50061 Section 6- Insulation coordination, lighting protection and grounding, Code for design of overhead power lines up to 66kV

Chapter 4
Hazardous Area Classification for International Oilfield Surface Facilities

Hazardous area classification provides a demonstrable methodology to identify areas in oil field where need to prevent fire and explosion with the potential risk of explosive atmosphere, and sources of ignition, and it is the significant guideline for oil field development. Hazardous Area Classification is important for international oilfield surface facilities, and it is not only the basis of Hazard and Operability Research (HAZOP) for the overall layout of the process in oilfield, but also provides necessary technical support for equipment selection and procurement. Due to the different developmental and technology level of petroleum industry, the standards and norms for Hazardous Area Classification are also different in the world.

At present industrial plant, API RP505 Recommended Practice for Classification of Locations for Electrical Installation at Petroleum Facilities Classified as Class I, Zone 0, Zone 1 and Zone 2 recommended by American petroleum institute and EI 15 Area Classification for Installations Handling Flammable Fluids recommended by UK Energy institute are the main standards used in the designing of international oilfield surface facilities. In some expansion projects, API RP 500 Recommended practice for classification of Locations for Electrical Installations at Petroleum Facilities Classified as Class I, Division 1 and Division 2 combined with EI 15 are adopted in special areas. Different standards adopted even in same project result in different areas classification, also with different solutions. Engineers must be familiar with the standard differences in these standards with professional experience when they execute the Hazardous Area Classification.

With regarding to Hazardous Area Classification, the standard for Petroleum Facilities in domestic is SY/T 6671 Recommended Practices for locations for electrical installations at petroleum facilities Classified as of Class I, Zone 0, Zone 1 and Zone 2. This domestic standard is based on API RP 505 as reference as recommended by the American Petroleum Institute.

Hereby, the areas classification difference with reference to the same project but with different standards is analyzed in order to compare the actual technical variance in different standards.

© Petroleum Industry Press 2022
K. Ma et al., *International Oilfield Surface Facilities: Safety Analysis for Electrical Design*, https://doi.org/10.1007/978-981-16-3104-7_4

This chapter introduces the commonly used in hazard area classification standards of oil field, followed by the classification example of typical hazardous areas based on different standards, then the results comparison will be analyzed, and final summary will be concluded. The names and versions of the standards mainly used for comparison in this chapter are as follows.

(1) SY/T 6671—2017 Recommended Practices for locations for electrical installations at petroleum facilities Classified as of Class I, Zone 0, Zone 1 and Zone 2;

(2) API RP 505: 2018 Recommended Practice for Classification of Locations for Electrical Installation at Petroleum Facilities Classified as Class I, Zone 0, Zone 1 and Zone 2;

(3) EI 15: 2015 Code of Safe Practice Part 15: Area Classification for Installations Handling Flammable Fluids;

(4) API RP 500: 2012 Recommended practice for classification of locations for Electrical Installations at Petroleum Facilities Classified as Class I, Division 1 and Division 2;

(5) IEC 60079-10-1: 2015 Explosive atmospheres—Part 10-1: Classification of areas—Explosive gas atmospheres;

(6) GB 3836.14-2014 Explosive atmospheres—Part 14: Classification of areas—Explosive gas atmosphere.

(7) GB 50058—2014 Code for design of electrical installations in explosive atmospheres.

1 General

The processing, handling and storage of oil and gas products in oilfield surface facilities inevitably results in the Leakage of flammable substances from pipelines, vessels, valves and pump. When flammable substances leak and mix with oxygen in the air, they will form an explosive gas mixture. If the concentration of the mixture is in the explosive concentration range (i.e. between the lower and upper explosive limits) and there is an electrical spark or hot surface present that is sufficient to ignite the explosive gas mixture, explosions and fires will be caused.

The purpose of the hazardous area classification is to guide the selection and installation of electrical and instrument equipment, to design the ventilation system and the location of fire protection equipment. The classification of hazardous area should not only ensure the safety and reliability of production facilities, but also avoid the excessive requirement of equipment investment cost caused by excessively increasing the level of hazard or expanding the scope of hazardous area, thus leading to the increase of total project investment.

Therefore, engineers should make good engineering judgments in conjunction with previous engineering experience during the preparation of hazardous area classification documents. Professional classification of hazardous areas is important significance for investment in oil field surface development projects, safety operation, prevention of fire and explosion accidents.

Hazardous areas shall be classified using the Zone system by U.S. National Electrical Code (NEC) in 1999 and by Canadian Electrical Code (CEC) in 1998. The principles and methods of hazardous area classification are described in detail in Section 505 of the NEC Code, published in 1999. As a stand-alone complete standard API RP 505 create the U.S. version of the Zone system which similar to the International Electrotechnical Commission (IEC) and the European Committee for Standardization (CENELEC). Compared with the Class/Division systems commonly used in North America before, API RP 505 is more in line with the manufacturing and installation of equipment of European electrical manufacturers, while maintaining part of the National electrical code wiring methods and protection technology. While standards for hazardous area classification are evolving towards harmonization although different standards and codes are different in the specific application.

2 Common Standards for Hazardous Areas Classification

Due to the different histories of oil industry development and the differences in oil and gas extraction technology between countries around the world, the Western countries have established an early understanding of the oil industry. The American Petroleum Institute (API) is one of the more complete and influential organizations in developing these norms and standards. American Petroleum Institute (API), the British Energy Association (Energy Institute, EI), National Electrical Code (National Electrical Code, NEC), International Electro technical Commission (IEC), etc. are also professional institutes.

In oilfield surface engineering, the following eight standards and codes are common adopted to hazardous areas classification.

(1) API RP 500 Recommended practice for classification of locations for Electrical Installations at Petroleum Facilities Classified as Class I, Division 1 and Division 2;

(2) SY/T 6671—2017 Recommended Practices for locations for electrical installations at petroleum facilities Classified as of Class I, Zone 0, Zone 1 and Zone 2;

(3) API RP 505 Recommended practice for classification of locations for Electrical Installations at Petroleum Facilities Classified as Class I, Zone 0, Zone1, and Zone 2;

(4) National Electrical Codes;

(5) EI 15 Model Code of Safe Practice—Part 15: Area Classification for Installations Handling Flammable Fluids;

(6) IEC 60,079–10-1 Explosive atmospheres—Part 10-1: Classification of areas— Explosive gas atmospheres;

(7)   GB 3836.14–2014 Explosive atmospheres—Part 14:Classification of areas—Explosive gas atmosphere
(8)   GB 50,058—2014 Code for design of electrical installations in explosive atmospheres.

API RP 500 and API RP 505 are mostly used in projects contracted by North America and American petroleum Companies. EI 15 and IEC 60,079-10 are adopted in the European region and by European oil companies. API RP 505 formulated by American Petroleum Institute and EI 15 formulated by British Energy Association have been widely used in the classification of dangerous zones. Domestic SY/T 6671-2017 is references to API RP 505.

2.1 Framework of API RP 505

API RP 505 is applicable for petroleum refineries, onshore and offshore fixed platforms, mobile offshore drilling platforms and pipeline transmission areas. API RP 505:2018 contains 14 chapters and 6 informative appendices as the following.

1.   Scope
2.   Normative References
3.   Terms, Definitions, and Acronyms
4.   Basic Conditions for a Fire or Explosion
5.   Flammable and Combustible Liquids, Gases, and Vapors
6.   Classification Criteria
7.   Extent of a Classified Location
8.   Recommendations for Determining Degree and Extent of Classified Locations—Common Applications
9.   Recommendations for Determining Degree and Extent of Classified Locations in Petroleum Refineries
10. Recommendations for Determining Degree and Extent of Classified Locations at Drilling Rigs and Production Facilities on Land and on Marine Fixed Platforms
11. Recommendations for Determining Degree and Extent of Classified Locations on Mobile Offshore Drilling Units (MODUs)
12. Recommendations for Determining Degree and Extent of Classified Locations at Drilling Rigs and Production Facilities on Floating Production Units
13. Reserved for Future Use
14. Recommendations for Determining Degree and Extent of Classified Locations at Petroleum Pipeline Transportation Facilities.

Annex A (informative) Sample Calculation to Achieve Adequate Ventilation of an Enclosed Area by Natural Means Using Eqs. 1 and 2.
Annex B (informative) Calculation of Minimum Air Introduction Rate to Achieve Adequate Ventilation Using Fugitive Emissions.

Annex C (normative) Symbols for Denoting Class I, Zone 0, Zone 1, and Zone 2 Hazardous (Classified) Areas.

Annex D (informative) An Alternate Method for Area Classification.

Annex E (informative) Procedure for Classifying Locations [1].

The oilfield surface engineering mainly refers to the content of Chaps. 10 and 14.

2.2 Framework of EI 15

EI 15 standard applies to virtually any location where there is a potential source of hazardous releases of flammable fluid substances. EI 15 contains 4 chapters and 8 informative appendices as the following.

1. Introduction
2. The technique of area classification
3. The point source approach for classification of individual sources of release
4. Effect of ventilation on area classification
5. Annex A classification and categorization of petroleum and flammable fluids
6. Annex B Area classification for hydrogen
7. Annex C Calculation of hazard radii
8. Annex D The direct example approach for classification of common facilities in open areas
9. Annex E Small-scale operations (laboratories and pilot plants)
10. Annex F Background and examples on ventilation of enclosure areas, releases within enclosed areas and associated external hazardous areas
11. Annex G Glossary
12. Annex H References [2].

2.3 Comparison on the Knowledge Framework of API RP 505 and EI 15

From the content structure of the two standards, the API RP505 standard application process is primarily based on a typical direct example approach. The EI 15 standard has both direct example method and point source analysis method in its application. It is important to note that the API RP505 standard describes the point source analysis in its Annex D (informative) An Alternate Method for Area Classification is similar to EI 15, which the three main steps of the hazard radius is used to classify the hazard area, point source identification, volatilization class determination, and areas classification. It is mentioned in the annotation of Appendix D that this method refers to EI 15, from which it can be seen that there is a relationship between the two standards, and then API RP505 standard is effectively supplemented and extended for the risk area classification method (Table 1).

Table 1 Comparison of the application scope and classification methods of API RP 505, EI 15 and IEC 60,079-10-1

Country/region	Standards	Application	Classification methods
North American	API RP 505,500	Places where flammable gases and vapours are present and associated petrochemical installations	Basic principles, direct example method and point source analysis
UK & Europe	EI Code Part 15	Places where flammable gases and vapours are present, including the entire plant area	Basic principles, direct example method, point source analysis and risk-based analysis
Europe	IEC 60,079-10-1	Places where flammable gases and vapours are present	Basic principles and direct example method

2.4 Hazard Area Classification Procedure

Hazardous area classification is usually studied in terms of the following considerations.

(1) What are the flammable substances that may be present?
(2) What are the physicochemical properties of each flammable substance?
(3) How the explosive atmosphere is formed by the release of the potential sources?
(4) Determine the pressure and temperature for routine system operation.
(5) Determine the ventilation conditions at site.
(6) Determine the concentration of flammable gases according to the diffusion conditions (ventilation level).
(7) Analyze the probability of occurrence of flammable substance release sites.
(8) Hazardous area classification based on above principle.

Based on the above analysis and data collection, a potential source release analysis table can be formed, refer to Table 2 Hazardous Source Release Analysis Table. After the Hazardous Source Release Analysis Table is formed, the hazardous range is divided according to the relevant specifications for hazardous area classification and finally it is shown on the hazardous area classification drawing (Tables 3 and 4).

3   EI 15 Classification of Hazardous Areas

The latest version of IEC 60,079-10-1 (2015 edition) on the classification of the explosion hazard zone is completely inconsistent with the old version, the earlier version is similar to API 505 and the China national standard, shows a typical equipment, area of the hazard zone radius, while the new version completely changed the basis and process of classification, and EI 15 is similar. The following EI 15 on the

Table 2 Hazardous source release analysis table—SOURCE OF RELEASE

Item	SOURCE OF RELEASE				
	Tag number	Description/location	Point of release sources	Grade of release	Release frequency level
Oil processing facilities					
1	MA-11210A/B	Inlet manifold	Flange leakage, accidental leakage	Secondary	Level 1
2	PR-11220/30/40/50	Pig receiver	Release during pigging operation	Primary	Level 1

Table 3 Hazardous source release analysis table—FLAMMABLE LIQUID

FLAMMABLE LIQUID					
Flammable material	Fluid class	Fluid category	Operating temp.& pressure		State
			(°C)	(MP ag)	
Crude oil	Class I	B	25–40	0.55	G/L
Crude oil	Class I	B	25–40	0.55	G/L

Table 4 Hazardous Source release analysis table—VENTILATION and HAZARDOUS AREA

VENTILATION		HAZARDOUS AREA			
Type	Degree	Zone type (0/1/2)	Zone extent		EI-15 code reference
			R1	R2	
N	A	2	4.0	4.0	C3, C4
N	A	1	3.0	3.0	3.6, C3, C4
		2	4.0	5.0	

explosion hazard zone classification process and some typical data to illustrate, as a simple guide to the classification of the explosion hazard zone.

3.1 Basic Processes for Hazardous Area Classification

The basic process for classifying hazardous areas in EI 15 is as follows.

(1) Determine whether the amount of flammable material exceeds Table 1.1 of the EI 15 standard, if not, classification is generally not required.

(2) If there is no possibility of leakage, there is no need to divide.

(3)   Classification of flammable materials.
(4)   Different classification method is selected depending on the type of process facility.

The first two steps in the basic process are primarily for small-scale facilities, such as laboratories, and the last two steps is considered for oilfield surface projects.

The flammables handled in the process can be classified according to EI 15 standard Table 1.3 and A1, which defines five flammable fluid media: A, B, C, G(i) and G(ii).

Depending on the type of oil and gas treatment facility, different hazardous area classification methods are followed after flammables are classified, and the EI 15 standard gives three classification methods: the direct example method, the point source method and the risk-based method [3].

3.2   Characteristics of Each Classification Method and Its Applicability in Oilfield Surface Engineering

3.2.1   Direct Example Method

The direct method is mainly adopted in storage tanks, road/rail/sea handling facilities, filling facilities, retail terminals, high volatility/high vapor pressure flammable storage facilities, drilling and well workover facilities, etc. The same or similar facilities meeting the conditions of the direct example method can be classified directly using the example of hazardous area classification given in the EI 15 standard.

The direct method is simple but with limited application scope, it is applicable to crude oil storage and loading stations in oilfield surface project.

3.2.2   Point Source Approach

The point source method can be used to hazardous area classification for facilities and equipment that are not suitable for the direct method. The point source method takes into account the radius of the hazardous area, the specific geometry of the equipment or facility, the shape factor, etc., to obtain a three-dimensional classification of the hazardous area classification.

The typical equipment to which the method applies is essentially the same as the main process equipment and facilities used in oilfield surface project, making the method be the most widely used in oilfield surface project.

3.2.3   Risk-Based Approach

The risk-based approach is used to determine the release point of a secondary release source when the release rate is uncertain, and it can also be used to adjust the release frequency and hazardous area radius to meet specific operating conditions.

Oilfield surface engineering process facilities are not highly intensive, mostly open-air installation and operation, ventilation conditions is better with no specific operating conditions, also the level of automation and monitoring of the oilfield DCS system could be able to reduce the workload of field inspections, since the point source method of hazard classification is sufficient to meet the requirements in most of cases, the risk-based method is generally not common used in oilfield surface engineering [4].

3.3   Scope of Application of the Point Source Method and Classification Procedure

3.3.1   Scope of Application

The point source method is mainly applicable to release sources of pumps, compressors, vent pipes, drainage of equipment, sampling points, piping systems, receivers and dispensers, and contaminated-oil basin to classify the hazardous area. The EI 15 standard lists these typical sources of releases which also are the main process equipment and facilities of the project.

3.3.2   General Steps and Content of the Point Source Approach

The point source approach to hazardous area classification is with four steps: identify the release source, determine the release source class and media classification, delineate the hazardous area category and determine the hazardous area radius, and determine the three-dimensional area of the hazardous area.

Release sources can be divided into three types, continuous release sources, primary release sources and secondary release sources. Long-term release and short-term release sources with a higher release frequency, or release sources with an annual release time of more than 1000 h are continuous release sources. Sources that are released periodically or occasionally during normal operation, or with an annual release time between 10 and 1000 h, are primary release sources. Sources that do not release during normal operation, that release only occasionally and for short periods of time, or that release for less than 10 h per year, are secondary release sources.

The classification of media should be based on the results of the physical properties of crude oil, wet gas and dry gas for specific engineering applications, or according to the general classification of EI 15 when conditions are not available, i.e. B for stable crude oil and G(i) for associated gas.

The hazardous area category relate to the level of the release source, the duration of the flammable environment and the intensity of the ventilation. In restricted outdoor conditions, ZONE 0 for areas with continuous sources and ZONE 1 for areas with primary sources, ZONE 2 for areas with secondary sources.

The hazard zone radius is determined by taking into account the operating pressure, the frequency of human exposure, the average number of release sources, the average probability of possible ignition, the flammable gas release level, the release frequency level and the release aperture.

3.3.3   Determination of the Extent of the Hazard Zone

When determining the radius of the hazard zone, EI 15 standard should firstly determine the release frequency level, then determine the release aperture, check the table according to different working conditions to obtain the hazard zone radius R1 and ground hazard zone radius R2, and finally consider the shape factor according to the height of the release source to obtain the three-dimensional area of the hazard zone.

1)   Release frequency level of a single release source per year. The annual release frequency levels for individual release sources shall be derived by the following procedure:

(1)   Determine P_{occ}, i.e. the probability that a person will be exposed to at least one potential release source. This shall be calculated by dividing the total number of hours per year that a single person is exposed to the hazardous area by the total number of hours per year (8760 h).

(2)   Determine N_{range}, i.e., the average number of secondary release sources. Typical values for N_{range} are given in the EI 15 standard: General inspection at open sites, $N_{range} = 1$; General inspection on compact station site, $N_{range} = 5$; inspection on area with a large number of release sources, $N_{range} = 30$.

(3)   Determination of E_{xp}, i.e. probability of exposure, $E_{xp} = P_{occ} * N_{range}$.

(4)   Determine the P_{ign}, i.e. the probability of ignition. EI 15 standard gives a typical value for P_{ign}: a controlled ignition source, $P_{ign} = 0.003$; weak ignition source, $P_{ign} = 0.01$; medium ignition source, $P_{ign} = 0.1$; strong ignition source, $P_{ign} = 1$. A land-based ignition source with a fire detection system is defined as a controlled ignition source; Typically ignition sources in Zone 2 are defined as weak ignition sources; ignition sources associated with traffic, substations, buildings, etc. are defined as moderate ignition sources; fire heaters, torches, etc. are defined as strong ignition sources.

(5)   Determine the release frequency level. EI 15 standard is based on E_{xp} calculations and P_{ign} values, which are derived from Fig. 1 (Refer to EI 15 Fig. C2). The figure is derived based on IR (individual risk) not greater than 10^{-5}.

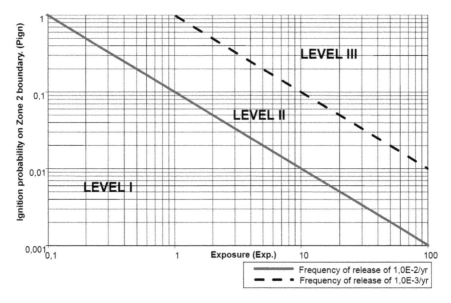

Fig. 1 Release frequencies—LEVEL

The EI 15 standard classifies release frequencies into three classes according to the following criteria.

- LEVEL I: Individual release sources with a release frequency greater than 10^{-2}/a.
- LEVEL II: Individual release sources with a release frequency of 10^{-2}/a–10-3/a.
- LEVEL III: The release frequency of a single release source is in the range of 10^{-3}/a–10^{-4}/a.

In EI 15 standard, LEVEL I is generally chosen for ZONE 2.

2) Release aperture

Equipment manufacturer's data should be prioritized used for release aperture. When actual data is not available, EI 15 standard defines equivalent release apertures for each release source based on different release frequency levels and release source types, as shown in Table 5 (Refer to EI 15 Table C13).

3) Scope of hazardous area

The EI 15 standard considers a combination of media type, source release frequency level, equivalent release aperture size and release pressure are developed in Table C4, which gives the hazard area radius R1 and the ground hazard area radius R2, respectively. See Table 6.

Taking into account the shape factor after R1 and R2 are confirmed, based on the relationship between the height of the release source above the ground, H, and R1, three cases are divided in EI 15 standard.

Table 5 Equivalent release apertures for typical release sources

Hole size (mm)

Equipment type	LEVEL I Greater than 10E-2/release source-yr	LEVEL II 10E-2–10E-3/release source-yr	LEVEL III 10E-3–10E-4/release source-yr
Single seal with throttle bush	2	5	10
Double seal	1	2	10
Reciprocating pump	2	10	20
Centrifugal compressor	1	5	30
Reciprocating compressor	2	10	30
Flanges	1	1	5
Vales	1	2	10

Notes

1. At the LEVEL I release frequency, for single seal centrifugal pumps with a throttle bush, use LEVEL II equivalent hole size
2. It is assumed that smaller equivalent hole sizes for valves > 80 mm diameter(when compared to valves < 80 mm diameter)are due to a higher mechanical integrity of the piping system, which will result in a lower failure frequency
3. Assumed LEVEL III failures are mainly due to the pump/ compressor sets and are generally independent of sealing arrangements

(1) $H > R_1 + 1$ m, no hazardous areas above ground.
(2) $R_1 + 1$ m $> H > 1$ m, a hazardous area on the ground, with a radius of R2 and a height of 1 m.
(3) $H \leq 1$ m, hazardous area on the ground, area radius R2, height 1 m [5].

4 Comparison of Typical Hazardous Area Classification

4.1 Subject of the Study and Operating Environment

In order to compare the different distance of the hazardous area caused by different standard, the following is a study of an export pump for transporting crude oil in a CPF station of an oil field in Iraq, with an installation height H equal to 1 m, crude oil fluid category B, and an operating pressure of 5.4 MPa. The pump is an API standard pump with single-layer seal and throttle bushing. In Consideration of the pump seal leakage, API RP 505 and EI 15 standards are used independently for the comparison of hazardous area classification ranges.

Table 6 Hazard radius R1 and R2

Fluid category	Release pressure (bar(a))	Hazard radius R1 (m) Release hole diameter				Hazard radius R1 (m) Release hole diameter			
		1 mm	2 mm	5 mm	10 mm	1 mm	2 mm	5 mm	10 mm
A	5^4	2	4	8	14	2	4	16	40
	10	2.5	4	9	16	2.5	4.5	20	50
	50	2.5	5	11	20	3	5.5	20	50
	100	2.5	5	11	22	3	6	20	50
B	5	2	4	8	14	2	4	14	40
	10	2	4	9	16	2.5	4	16	40
	50	2	4	10	19	2.5	5	17	40
	100	2	4	10	20	3	5	17	40
C	5	2	4	8	14	2.5	4	20	50
	10	2.5	4.5	9	17	2.5	4.5	21	50
	50	2.5	5	11	21	3	5.5	21	50
	100	2.5	5	12	22	3	6	21	50
G(i)	5	<1	<1	<1	1.5	<1	<1	1	2
	10	<1	<1	1	2	<1	<1	1.5	3
	50	<1	1	2.5	5	<1	1.5	3.5	7
	100	<1	1.5	4	7	1	2	5	11

(continued)

Table 6 (continued)

Fluid category	Release pressure (bar(a))	Hazard radius R1 (m) Release hole diameter				Hazard radius R1 (m) Release hole diameter			
		1 mm	2 mm	5 mm	10 mm	1 mm	2 mm	5 mm	10 mm
G(ii)	5	<1	<1	1.5	3	<1	<1	2	3
	10	<1	1	2	4	<1	1	2.5	5
	50	<1	2	4	8	1	2	4	11
	100	1	2	6	11	2	3	6	14
LNG	1.5	2.5	3	6	10	2	3	7	30
	5	3	5	10	17	2	4	11	40
	10	3	55	10	20	2.5	4.5	13	37.5

Notes

1. At the fluid storage temperatures of 20 °C the nominal discharge pressure of 5 bar(a) is below the saturated vapour pressure of Fluid category A. The saturated vapour pressure (6.8 bar(a)) was used to calculate the discharge rate and dispersion

2. Distances of LFL for LNG releases at 5 m height. These distances have been modeled as methane, with typical LNG compositions varying between 93–90%. Typical rundown, storage and loading temperatures for LNG are in the range −170 to −160 °C; therefore releases from a storage temperature of −165 °C have been modeled

3. No data are available for gasoline blends with ethanol; however, for blends with small quantities of ethanol, these could be treated as category C. It is recommended that modeling is carried out

4. Release pressure should be taken as the maximum allowable operating pressure

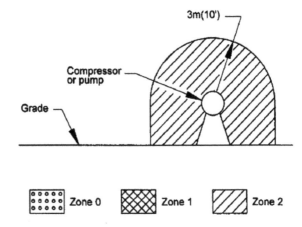

Fig. 2 Compressors and pumps in an adequately ventilated unconfined area [6]

4.2 API RP 505 Direct Example Method

4.2.1 General Application

From Fig. 2, it can be seen that the hazardous area is divided into R = 3 m [6].

4.2.2 Oil Pipeline Transmission Facility Areas

As shown in Fig. 3, the hazardous area is divided into R = 7.5 m, L = 15 m, D = 0.6 m [7].

4.3 API RP 505 Appendix D: Proxy Methods for Zoning

From Fig. 4, it can be seen that the hazardous area delineation range D1 = 7.5 m, H1 = 7.5 m, D2 = 7.5 m, H2 = 7.5, D3 = 7.5 m, H3 = 0.6 m [8] (See Table 7).

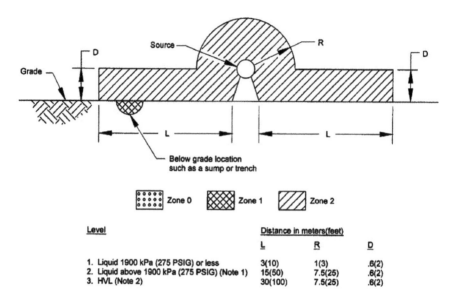

Level	Distance in meters(feet)		
	L	**R**	**D**
1. Liquid 1900 kPa (275 PSIG) or less	3(10)	1(3)	.6(2)
2. Liquid above 1900 kPa (275 PSIG) (Note 1)	15(50)	7.5(25)	.6(2)
3. HVL (Note 2)	30(100)	7.5(25)	.6(2)

Fig. 3 Compressors and pumps for outdoor transport of flammable or highly volatile liquids [7]

4.4 EI 15 Point Source Approach Based on Risk Analysis

Sources of release in the pump installation area are generally considered the pump seals, flanges where the pump connects to the line, valves, filters on the line, and the venting system.

According to the working conditions of the field operator and maintainer, they are mainly exposed to Zone 2 during their working hours, check Table 8 Exposure Parameters Table and Table 9 Probability of Ignition Table, the P_{occ} value is 0.13, the average N_{range} of the number of secondary release sources in the area is 11.7, then the exposure parameter $E_{XP} = P_{occ} * N_{range} = 1.521$. The ignition probability of P_{ign} takes 0.019, and the release frequency level is LEVEL I, and the release frequency is higher than 1.0–2 per year, according to Fig. 1.

Based on the data in Table 5, the release equivalent radius is confirmed, For a single seal with throttle bushing API standard pump, the equivalent radius of 0.1 times the shaft diameter, such as external pump shaft diameter of 80 mm, equivalent radius of 80 * 0.1 = 8 mm, should be noted that in general the pump seal leakage equivalent aperture data is usually provided by the manufacturer, in the absence of manufacturer data support, refer to Table 5 release frequency equivalent aperture table for equivalent aperture calculation.

Checking the hazardous radius table (R_1) and (R_2) in Table 4: $R_1 = 10$ m and $R_2 = 17$ m [9].

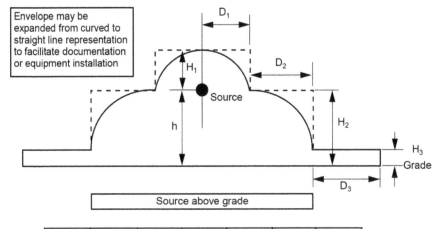

Envelope may be expanded from curved to straight line representation to facilitate documentation or equipment installation

Source above grade

Hazard Radius m (ft)	D_1 m (ft)	H_1 m (ft)	D_2 m (ft)	H_2 m (ft)	D_3 m (ft)	H_3 m (ft)
1 (3)	1 (3)	1 (3)	0 (0)	NA	2 (7)	0.5 (1.5)
1.5 (5)	1.5 (5)	1.5 (5)	0 (0)	NA	3 (10)	0.5 (1.5)
3 (10)	3 (10)	3 (10)	0 (0)	NA	3 (10)	0.6 (2)
5 (15)	5 (15)	5 (15)	0 (0)	NA	3 (10)	0.6 (2)
7.5 (25)	6 (20)	6 (20)	1.5 (5)	3 (10)	6 (20)	0.6 (2)
15 (50)	7.5 (25)	7.5 (25)	7.5 (25)	7.5 (25)	7.5 (25)	0.6 (2)
30 (100)	7.5 (25)	7.5 (25)	7.5 (25)	7.5 (25)	7.5 (25)	0.6 (2)

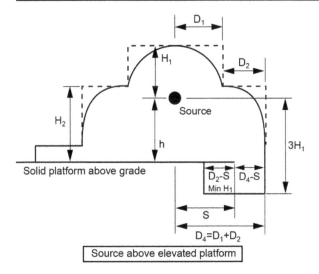

Solid platform above grade

Source above elevated platform

Fig. 4 Adequately ventilated process area with heavier-than-air gas or vapor Source Located Near or Above Grade(refer to API RP 505 Fig. D3)

Table 7 Pumps handling heavier-than-air gases or vapors located in non-enclosed, adequately ventilated process areas

Pump	Category	Low flow <100 gpm			Medium flow 100–500 gpm			High flow >500 gpm			Pump flow rate / Seal chamber pressure
Low = <100 psig, High = >500 psig		Low pressure	Medium Pressure	High pressure	Low pressure	Medium Pressure	High pressure	Low pressure	Medium Pressure	High pressure	Hazard radius (ft)
Standard pump	1	15	25	50	25	50	100	25	50	100	Hazard radius (ft)
	2	10	15	25	10	25	50	15	25	50	
	3	3	10	15	5	10	25	15	15	25	
High technology low seal emission pump	1	5	10	15	5	10	25	10	10	25	Hazard radius (ft)
	2	3	5	10	3	5	10	5	10	10	
	3	3	3	5	3	3	5	5	5	10	

Table 8 Exposure parameters table

Work pattern				NO. of release sources within range				Exposure
				% of time		N_{range}		(Exp)
Average Hours/yr spent on site	Fraction of time on site spent within plant area	Hours/yr spent on site within radius of plant area	P_{occ} Fraction of total time per yr spent within plant area	Open plant 1 source (%)	Congested plant 5 sources (%)	Many release sources 30 sources (%)	Average number of sources in range during time on site	P_{occ} * N_{range}
1920	1	1920	0.220	0	0	100	30	6.6
1920	1	1920	0.220	20	30	50	16.7	3.7
1920	1	1920	0.220	20	50	30	11.7	2.6
1920	1	1920	0.220	50	30	20	8	1.8
1920	1	1920	0.220	100	0	0	1	0.2
1920	0.6	1152	0.130	0	0	100	30	3.9
1920	0.6	1152	0.130	20	30	50	16.7	2.2
1920	0.6	1152	0.130	20	50	30	11.7	1.5
1920	0.6	1152	0.130	50	30	20	8	1.0
1920	0.6	1152	0.130	100	0	0	1	0.13
1920	0.25	480	0.055	0	0	100	30	1.65
1920	0.25	480	0.055	20	30	50	16.7	0.92
1920	0.25	480	0.055	20	50	30	11.7	0.64
1920	0.25	480	0.055	50	30	20	8	0.44
1920	0.25	480	0.055	100	0	0	1	0.06
1920	0.125	240	0.028	0	0	100	30	0.8
1920	0.125	240	0.028	20	30	50	16.7	0.5
1920	0.125	240	0.028	20	50	30	11.7	0.3
1920	0.125	240	0.028	50	30	20	8	0.2
1920	0.125	240	0.028	100	0	0	1	0.03

5 Data Collation and Analysis for Hazardous Area Classification

5.1 Hazard Scoping Data Collation and Analysis

The data from the hazard area scoping by the methods used in Sects. 2–4 above are collated and equivalently converted to R_1 and R_2 values as shown in Table 10 Hazard Area Scoping Radius Comparison.

From the comparative table of hazardous area radii as shown in Table 10, the application of the API RP 505 Annex D method for hazardous area classification gives the largest hazard radius range; the application of the API RP 505 direct example method for general application gives the smallest hazard radius range; and the EI 15 point source method based on risk analysis yields a hazard radius range in the middle of the two methods.

Table 9 Probability of Ignition Table

Percentage of time worker spends in areas with following ignition sources at the plant boundary				Pign
Strong (%)	Medium (%)	Weak (%)	Controlled (%)	
100	0	0	0	1.000
40	40	20	0	0.442
20	40	40	0	0.244
10	50	40	0	0.154
0	100	0	0	0.100
0	60	40	0	0.064
0	50	50	0	0.055
0	40	60	0	0.046
0	10	90	0	0.019
0	0	100	0	0.010
0	0	90	10	0.009
0	0	50	50	0.007
0	0	0	100	0.003

Table 10 Hazardous area classification radius comparison table

Standard	Method description	R_1(m)	R_2(m)
API RP 505	Sample general application	3	3
API RP 505	Oil pipeline transmission facilities area	7.5	15
API RP 505	Annex D zoning proxies	7.5	30
EI 15	source approach based on risk analysis	10	17

Note R_1 is the release radius in the vertical direction and R_2 is the release radius in the horizontal direction after the influence of external shape factors

5.2   Hazard Scoping Data Analysis

The hazard range radius obtained from the general application of API RP 505 direct example method does not take into account the volatility of flammable substances, release rates and operating pressure, and does not accurately reflect the actual operating conditions of various types of oil and gas treatment facilities, and the 3 m hazardous radius R2 reference data provided is relatively conservative.

The application of API RP 505 Direct Example Method to the area of a petroleum pipeline transmission facility takes into account the pressure level, volatility of flammable substances and subdivides the hazard range radius into three categories: hazard range radius for liquid vapour pressures less than or equal to 1900 kPa, hazard range radius for liquid vapour pressures greater than 1900 kPa, and hazard range radius for highly volatile liquids. The general application of the example method has been subdivided by pressure class compared to API RP 505, which extends the value of the hazard radius R2 to 15 m.

The API RP 505 Annex D—Area Classification Proxy Method is actually quite similar to the point source method application mentioned in EI 15, where the point release source concept is first introduced and the fluid class classification is described, which in turn determines the hazard range radius.

In the implementation of the hazardous area classification, the release rate matrix is provided to determine the hazardous radius according to the volatile category and the release rate, which can find out the possible hazardous radius of each type of fluid at different release rates. For the international oilfield pumps, which belong to the matrix category 1, high release rate, the value of R2 is further extended to 30 m. 30 m hazardous area radius should be avoided as far as possible in the development and design of oil field surface project applications, because the formation of a large Zone 2 has heavy impact on the selection and layout of equipment in industrial plant and the neighboring areas.

The point source method of EI 15 based on risk analysis is based on considering the probability of the exposure of field operators to the risk of ignition of flammable substances within the range of the potential hazardous area, and then establish the release frequency level, according to the release frequency level to calculate the equivalent release aperture, after checking the defined table to establish the hazard radius. The method is fully considered the factors affecting the potential hazardous release sources, and for the external transmission pump in this study, the hazard radius R2 value due to seal leakage is 17 m. Compared with the 30 m radius area formed by the API RP 505 Annex D Zoning Proxies method, the hazard area radius is greatly reduced to an acceptable range.

By comparing the hazardous area classification in adopting API RP 505 standard and EI 15 standard to study the external transmission pump in an oilfield CPF station due to potential seal leakage, it can be concluded that when detailed physical properties of combustible fluids and relevant process handling parameters are established, the hazard range radius delineated with EI 15 standard is smaller than that using API RP 505 standard, which has better engineering design application and practical guidance significance in the oilfield surface engineering development process.

The direct example method of API RP 505 standard for pump with flammable fluids has a pressure classification point of 1.9 MPa, resulting in a 15 m radius range for pressures higher than 1.9 MPa and a 3 m radius range for pressures less than 1.9 MPa. Even with the application of Annex D—An Alternate Method for Area Classification, the pressure division is only with limited class of less than 0.7 MPa low pressure, 0.7–3.4 MPa medium pressure and higher than 3.4 MPa high pressure of the three levels. In the application of EI 15 standard, according to the fluid category, the pressure classification is divided into four levels: 0.5, 1, 5 and 10 MPa, which is more accurate than that of the API RP505 standard. Therefore, EI 15 standard provides more reference data than the API RP 505 standard when the source of the hazard is within the defined pressure range. With reference to the high-pressure facilities, the EI 15 standard of 10 MPa level is also much higher than the API RP 505 standard of 3.4 MPa, therefore, it is more convenient to adopt the EI 15 standard for the classification of the hazardous area of the oil and gas with processing high-pressure system.

In the oilfield surface facilities engineering, the classification of hazardous areas should also be based on the actual condition of the project and the reasonable selection of compliance criteria. The locations, historical and cultural background, the scope of bulk material procurement and the location of equipment suppliers and etc. have to be taken into consideration in order to define which standard is govern for this project. The renovation or expansion project should follow the original design standards as much as possible to maintain the uniformity of the project. If the project standards are specified by the ITB from client, it has to be implemented in accordance with the client's requirements.

In combination, the legend method together with the Alternate Method for Area Classification of API RP 505 Annex D, provides a method in the hazardous area classification process. Typical diagram according legend method can be used as engineering drawing documents. For equipment and systems with high conveying pressures, it recommend to use the alternate method for Area Classification to further divide and find a reasonable range of division.

The application of the EI 15 standard requires the cooperation of several disciplines, such as process, equipment, HVAC, instrumentation and environment, etc. The validity of this large amount of input data needs to be justified and be confirmed by professional engineers. The hazard area classification should be partially modified as vendor data is submitted in the later stage of the project.

Each standard has its own limitations from the above comparison, and it is unrealistic to require common procedure from one particular standard for all kinds of situations. Therefore, in the process of standard application, engineers who execute hazardous area classification should not only be familiar with the process conditions and scope of application of each standard, but also should take the existing project experience into consideration, combine with past engineering practice to make comprehensive judgment, and reasonably classify the hazardous area, and to ensure the safety and economy of the project.

References

1. API RP 505: 2018 Recommended Practice for Classification of Locations for Electrical Installation at Petroleum Facilities Classified as Class I, Zone 0, Zone 1 and Zone 2
2. EI 15: 2015 Code of Safe Practice Part 15: Area Classification for Installations Handling Flammable Fluids
3. EI 15: 2015 Code of Safe Practice Part 15:Area Classification for Installations Handling Flammable Fluids, Section 1
4. EI 15: 2015 Code of Safe Practice Part 15: Area Classification for Installations Handling Flammable Fluids, Section 2
5. EI 15: 2015 Code of Safe Practice Part 15: Area Classification for Installations Handling Flammable Fluids, Section 3
6. API RP 505: 2018 Recommended Practice for Classification of Locations for Electrical Installation at Petroleum Facilities Classified as Class I, Zone 0, Zone 1 and Zone 2, Section 10

7. API RP 505: 2018 Recommended Practice for Classification of Locations for Electrical Installation at Petroleum Facilities Classified as Class I, Zone 0, Zone 1 and Zone 2, Section 14
8. API RP 505: 2018 Recommended Practice for Classification of Locations for Electrical Installation at Petroleum Facilities Classified as Class I, Zone 0, Zone 1 and Zone 2, Annex D
9. EI 15: 2015 Code of Safe Practice Part 15: Area Classification for Installations Handling Flammable Fluids, Annex C

Appendix

Example of ESSID and ESTOS worksheets during electrical engineering of one of international oil field development project (See Figs. A.1, A.2, A.3, A.4, A.5, A.6 and Tables A.1, A.2, A.3, A.4, A.5, A.6, A.7, A.8, A.9, A.10, A.11, A.12).

© Petroleum Industry Press 2022
K. Ma et al., *International Oilfield Surface Facilities: Safety Analysis for Electrical Design*, https://doi.org/10.1007/978-981-16-3104-7

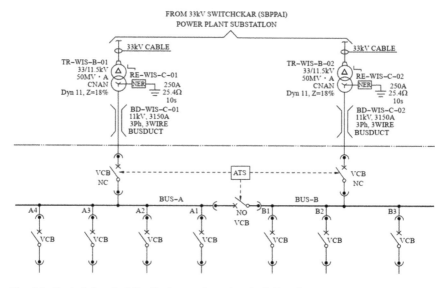

Fig. A.1 Typical electrical distribution configuration single line diagram

Equipment:

1. 33 kV cable from 33 kV Power Plant Substation to 33/11.5 kV Transformer
2. 33/11.5 kV 50MVA Transformer
3. 250 A NER
4. 11 kV 3150 A Bus duct.

Table A.1 Typical ESSIS worksheet for general system

Parameter	Guideword	Cause	Consequences	Recommendation	Action party	Remarks/notes
Tagging	Undefined system	Equipment tag is not following PTS standard	Confusion during operation and switching that could lead to operation error	EPC/OWNER to review the tagging system to ensure consistency during EPC stage	EPC/OWNER	NO ISSUE The tagging system is unique for total project development
Availability	No	Transformer with non-standard design might not be available Might be long lead time item for procurement	Transformer with non-standard design might not type tested	To check the availability of the transformer with OEM	EPC/OWNER	NO ISSUE EPC/OWNER confirms that the delivery of the transformer will have no schedule impact
Rating	Under	Transformer impedance is very high compared to standard transformer impedance	High impedance can cause unnecessary high voltage drop at the downstream switchgears	To reconfirm the impedance value for the transformer	EPC/OWNER	NO ISSUE EPC/OWNER has consulted with OEM on the matter
Voltage	Over	System voltage surges due to lightning	Damage to equipment	Surge protection to be provided at the incomer of the switchgears	EPC/OWNER	NO ISSUE Surge protection is provided at the incomer of the switchgears
Voltage	Under	Loss of power from Upstream feeder	Big voltage drop could lead to equipment failure	To develop a load shedding scheme	EPC/OWNER	NO ISSUE Provision for load shedding scheme has been provided at the IMCS/EICS
Protection and control	Inadequate	Ambient temperature at site is very high. No protection for oil flashpoint	Transformer might overheat and trip	To confirm the alarm/trip setting for the transformer oil flash point	EPC/OWNER	NO ISSUE The transformer ambient temperature design is 55 °C The setting for alarm is at 100 °C and will trip at 105 °C

(continued)

Table A.1 (continued)

Parameter	Guideword	Cause	Consequences	Recommendation	Action party	Remarks/notes
Protection and control	Inadequate	Differential protection cannot communicate and function with upstream switchgear	Differential protection relay of different type and connection	CT for differential to be identical in connection and communication method with the upstream feeder	EPC/OWNER	
Environment-humidity	High	Design of the busbars to accommodate for insulation and humidity	Flashover	To confirm the busbars design (distance between busbars, layers, insulation) and the heat dissipation data shall be included for the HVAC sizing calculation	EPC/OWNER	
Interlock	Inadequate	Difficult to access the CT/PT	Flashover	To confirm the design for the earthing of the PT/VT whether it is withdrawable type. To provide the special tools needed to access the PT/VT	EPC/OWNER	
Current	Over	Upstream fault, system short circuit	Damage to equipment	Protection coordination study to be conducted during EPC stage and completed prior to commissioning	EPC/OWNER	
Protection and control	No	The power supply of the space heater is down during emergency	Damage to equipment	To design the power supply of the switchgear space heater to be provided from the emergency switchgears for all the switchgears	EPC/OWNER	
Current	Over	Upstream fault, short circuit	Damage to switchgear	The short circuit study, load flow and motor starting study shall be finalised upon completion of data submission from OEM	EPC/OWNER	

(continued)

Table A.1 (continued)

Parameter	Guideword	Cause	Consequences	Recommendation	Action party	Remarks/notes
Protection and control	Inadequate	CT is not sensitive to detect fault	Protection might not pick up the fault	Selection of let through current and relay proposal shall be provided by switchgear VENDOR. All functional test for the protection to be conducted during FAT	EPC/OWNER	
Humidity	High	No safety switch and earth fault protection for motor space heater	Condensation on motor could lead to short circuit or motor failure	Space heater supply is provided with RCBO rated 30 mA. Safety switch is provided near the motors	EPC/OWNER	NO ISSUE
Current	Over	Large cables with improper installation	Cable cause mechanical damage due to fault and cable is not clamped	To confirm the cable installation method and the distance interval for the cable cleats	EPC/OWNER	
Protection and control	No	The differential relay between upstream and downstream switchgears not activated	Protection not able to pick up fault	The Class X CT shall be manufactured from the same manufacturer	EPC/OWNER	

(continued)

Table A.1 (continued)

Parameter	Guideword	Cause	Consequences	Recommendation	Action party	Remarks/notes
Protection and control	No	Absence of earth fault protection for outgoing feeder to external load centres	Trip at upstream	Earth fault protection is provided for the main incomer breaker of all downstream distribution boards	EPC/OWNER	<u>NO ISSUE</u> Provision has been made as below: If more than 100A, MCCB with LSIG is considered. If less than 100A, RCB0 is considered Distribution boards located more than 200 m away from the switchgear, earth fault protection shall be provided for the switchgear outgoing feeder also
Interlocking	Availability	Protection cannot be initialised after tripping and during faulty relay	Downstream loads are tripped	To design for latched type lock out relay (86) in an independent relay with bypass switch in case of relay failure	EPC/OWNER	

Table A.2 Typical ESTOS worksheet for general system

Elements	Potential hazards and/or operability issues	Causes	Key tasks (safety mitigation plan)	Recommendations	Action by
Transformer	Fire/explosion—others	Transformer fault or due to switching surges	N/A	To provide the Transformer bay with Fire Protection system and fire extinguisher Note: Project has already provided the required Fire Protection system and fire extinguisher at the transformer bays	EPC/OWNER
Cable/bus duct	Mechanical hazard—others	Blocked access	To create adequate access	Bus duct is bottom entry. To restudy the design of the bus duct routing with minimal joints and bends and consideration of access during construction and operation maintenance	EPC/OWNER
Transformer	Fire/explosion—flashover/checklist—visual inspection	Inadequate connection, hot spots	To ensure that ample separation/segregation is provided at the termination box based on the number of incomer cables	To ensure that ample separation/segregation is provided at the termination box based on the number of incomer cables To confirm with OEM for cable termination: 1. No. of cables 2. Size of cables 3. Installation method 4. Temperature/humidity/condensation	EPC/OWNER
Transformer	Mechanical Hazard—Fences	Transformer cannot be taken out for service or maintenance	To ensure that the transformer can be taken out for service	To consider the mechanical handling for the transformer to be taken out for service and maintenance	EPC/OWNER

(continued)

Table A.2 (continued)

Elements	Potential hazards and/or operability issues	Causes	Key tasks (safety mitigation plan)	Recommendations	Action by
Transformer	Mechanical hazard—trip over	Cables laid on the floor un arranged and unprotected	To ensure cables are arranged and protected adequately without blocking any access for maintenance	To ensure that cable routing near the equipment will not pose any tripping hazard to people working near the transformer and generator	EPC/OWNER
Transformer	Electrocution—direct contact	Terminal box too short to install the HV termination kit inside the cable compartment. Additional drop box might be required which will affect the IP and form of the switchgear	To provide adequate space for termination of cables	To confirm the cable compartment size and the gland plate of the cable compartment prior to procurement of switchgear to ensure no drop box is needed. To ensure that vertical support is provided adequately to support the weight of the incoming cables	EPC/OWNER
Cable/bus/bus duct	Mechanical hazard—others	Bus duct cracks	To support bus duct	To design bus duct support to adequately support the bus duct weight To design for flexible bus duct at both connection ends To design the bus duct plinth with that of the transformer for outdoor portion of bus duct	EPC/OWNER
Transformer	Fire/explosion—others	Spread of fire to all transformer bays through adjoining oil sump	To segregate oil sump of each transformer bay	To design for dedicated oil sump for each transformer with no connected drain pipe to other transformer bays To discuss this design with Civil discipline during revision of substation drawing	EPC/OWNER
Transformer	Fire/explosion—others	Transformer by collapses due to non-fire rated wall during fire	To uses fire rated wall as separation wall between transformers	To uses fire rated wall as separation wall between transformers Note: Project has already design transformer bay separation walls with 2-h fire rated walls	EPC/OWNER

(continued)

Table A.2 (continued)

Elements	Potential hazards and/or operability issues	Causes	Key tasks (safety mitigation plan)	Recommendations	Action by
Earthing system	Electrocution—proximity	For NER	To prevent condensation on the NER enclosure	To provide NER enclosure with sufficient ventilation to avoid condensation	EPC/OWNER
Transformer	Environment	Not enough visibility for transformer maintenance	To provide adequate lighting during normal and emergency	To design for availability of normal and emergency lighting for visibility during maintenance/service of the transformer	EPC/OWNER
Bus duct	Fire/explosion—flashover/checklist—visual Inspection	Water ingress and dust accumulation at the bus duct cause damage to the bus duct	To ensure bus duct is always clean and dry	To ensure that all penetration for the bus duct are properly sealed with prevention again water and dust ingress	

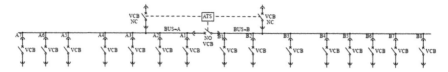

Fig. A.2 Typical MV switchgear single line diagram

Table A.3 ESSID—MV switchgear

Parameter	Guideword	Cause	Consequences	Recommendation	Action party	Remarks/notes
Tagging	Undefined system	Equipment tag is not following standard	Confusion during operation and switching that could lead to mal operation	Review the tagging system to ensure consistency during EPC stage.	All	Tagging
Voltage	Over	System voltage surges due to lightning	Damage to equipment	Surge protection to be provided at the incomer of the switchgears	EPC/OWNER	NO ISSUE Surge protection is provided at the incomer of the switchgears
Voltage	Under	Loss of power from upstream feeder	Big voltage drop could lead to equipment failure	To develop a load shedding scheme	EPC/OWNER	NO ISSUE Provision for load shedding scheme has been provided at the IMCS/EICS
Protection and control	Inadequate	Ambient temperature at site is very high. No protection for oil flashpoint	Transformer might overheat and trip	To confirm the alarm/trip setting for the transformer oil flash point	EPC/OWNER	NO ISSUE The transformer ambient temperature design is 55 °C The setting for alarm is at 100 °C and will trip at 105 °C
Protection and control	Inadequate	Differential protection cannot communicate and function with upstream switchgear	Differential protection relay of different type and connection	CT for differential to be identical in connection and communication method with the upstream feeder	EPC/OWNER	
Environment-humidity	High	Design of the busbars to accommodate for insulation and humidity	Flashover	To confirm the busbars design (distance between busbars, layers, insulation) and the heat dissipation data shall be included for the HVAC sizing calculation	EPC/OWNER	

(continued)

Table A.3 (continued)

Parameter	Guideword	Cause	Consequences	Recommendation	Action party	Remarks/notes
Interlock	Inadequate	Difficult to access the CT/PT	Flashover	To confirm the design for the earthing of the PT/VT whether it is withdrawable type To provide the special tools needed to access the PT/VT	EPC/OWNER	
Current	Over	Upstream fault, system short circuit	Damage to equipment	Protection coordination study to be conducted during EPC stage and completed prior to commissioning	EPC/OWNER	
Protection and control	No	The power supply of the space heater is down during emergency	Damage to equipment	To design the power supply of the switchgear space heater to be provided from the emergency switchgears for all the switchgears	EPC/OWNER	
Current	Over	Upstream fault, short circuit	Damage to switchgear	The short circuit study, load flow and motor starting study shall be finalised upon completion of data submission from OEM	EPC/OWNER	
Protection and control	Inadequate	CT is not sensitive to detect fault	Protection might not pick up the fault	Selection of let through current and relay proposal shall be provided by switchgear VENDOR All functional test for the protection to be conducted during FAT	EPC/OWNER	

(continued)

Table A.3 (continued)

Parameter	Guideword	Cause	Consequences	Recommendation	Action party	Remarks/notes
Humidity	High	No safety switch and earth fault protection for motor space heater	Condensation or motor could lead to short circuit or motor failure	Space heater supply is provided with RCBO rated 30 mA Safety switch is provided near the motors	EPC/OWNER	NO ISSUE
Protection and control	No	The differential relay between upstream and downstream switchgears not activated	Protection not able to pick up fault	The Class X CT shall be manufactured from the same manufacturer	EPC/OWNER	
Protection and control	No	Absence of earth fault protection for outgoing feeder to external load centres	Trip at upstream	Earth fault protection is provided for the main incomer breaker of all downstream distribution boards	EPC/OWNER	NO ISSUE Provision has been made as below: If more than 100 A, MCCB with LSIG is considered. If less than 100 A, RCB0 is considered Distribution boards located more than 200 m away from the switchgear, earth fault protection shall be provided for the switchgear outgoing feeder also
Interlocking	Availability	Protection cannot be initialised after tripping and during faulty relay	Downstream loads are tripped	To design for latched type lock out relay (86) in an independent relay with bypass switch in case of relay failure	EPC/OWNER	

Table A.4 ESATO-MV switchgear

Elements	Potential hazards and/or operability issues	Causes	Key tasks (safety mitigation plan)	Recommendations	Action by
Switchgear	Fire/explosion—flashover/checklist—visual inspection	Busbar loose connection, hot spots	N/A	To ensure that the Arc Flash capability is available for all switchgears	EPC/OWNER
Switchgear, transformer	Fire/explosion—flashover/checklist—visual inspection	Inadequate connection, hot spots	To ensure that ample separation/segregation is provided at the termination box based on the number of incomer cables	To ensure that ample separation/segregation is provided at the termination box based on the number of incomer cables. To confirm with OEM for cable termination: 1. No. of cables 2. Size of cables 3. Installation method 4. Temperature/ humidity/ condensation	EPC/OWNER
Switchgear	Mechanical hazard—blockage	Switchgear cannot be installed or taken out for service or maintenance	To ensure that the switchgear can be installed or taken out for service	To consider the largest panel size for material handling when sizing the substation doors	EPC/OWNER
Transformer Switchgear	Electrocution—direct contact	Terminal box too short to install the HV termination kit inside the cable compartment. Additional drop box might be required which will affect the IP and form of the switchgear	To provide adequate space for termination of cables	To confirm the cable compartment size and the gland plate of the cable compartment prior to procurement of switchgear to ensure no drop box is needed. To ensure that vertical support is provided adequately to support the weight of the incoming cables	EPC/OWNER
Switchgear	Electrocution—direct contact	Installing new cables for Phase 2 motors with Phase 1 motors running	To provide adequate space access for installing Phase 2 equipment without interrupting the operation of Phase 1 equipment	To ensure that adequate measures have been taken for phase 2 project cable installation without interrupting current operation	EPC/OWNER

(continued)

Table A.4 (continued)

Elements	Potential hazards and/or operability issues	Causes	Key tasks (safety mitigation plan)	Recommendations	Action by
Switchgear	Accessibility	Future switchgears blocked and not accessible	To provide adequate space access for installing Phase 2 equipment without interrupting the operation of Phase 1 equipment	To provide adequate space for future switchgear To provide cut out for future expansion panels and cover with metal plate To ensure that no blocking for operation and maintenance of the switchgear panels To ensure that the installation of the Phase 2 Project motor cables shall not be blocked by Phase 1 Project motor cables	EPC/OWNER
Switchgear	OI procedure—PTW	Improper switching	Indicators for switching condition for the circuit breaker	To develop the switching procedure and provide indication at the switchgears	EPC/OWNER
Switchgear	Flashover	Water ingress into switchgears	To prevent water ingress into switchgears	To design the HVAC system such that duct diffusers/blower fans are not installed directly on top of the switchgears	EPC/OWNER
Switchgear	Fire/explosion—flashover/checklist—visual inspection	No tagging at the back	To ensure adequate tagging	Tagging to be provided at both front and back sides of the switchgear	EPC/OWNER
Switchgear	Electrocution—direct contact	PT/VT cannot be accessed during maintenance	To ensure access to PT/VT is available	To design the switchgear with front access to PT/VT with no blockage during maintenance/servicing To provide all the special tools required to withdraw the PT/VT	EPC/OWNER
Protection Relay	OI procedure—work instruction	No indicator of fuse blown for incoming cable measurement system	To provide indication during fault	To design for physical indicator at switchgear panel during CT/PT fault	EPC/OWNER

(continued)

Table A.4 (continued)

Elements	Potential hazards and/or operability issues	Causes	Key tasks (safety mitigation plan)	Recommendations	Action by
Switchgear	Mechanical hazard—others	Switchgear cannot be accessed due to blockage by column	To ensure adequate space around the switchgear for maintenance and operation	To ensure that all switchgears are easily access and is not blocked To provide dummy panels for switchgears panels that is located adjacent to a structural column	

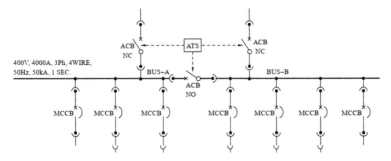

Fig. A.3 LV switchgear single line diagram

Table A.5 ESSID—LV switchgear

Parameter	Guideword	Cause	Consequences	Recommendation	Action party	Remarks/notes
Tagging	Undefined system	Equipment tag is not following PTS Standard	Confusion during operation and switching that could lead to operation error	EPC/OWNER to review the tagging system to ensure consistency during EPC stage	EPC/OWNER	NO ISSUE The tagging system is unique for total project development
Environment-humidity	High	Design of the busbars to accommodate for insulation and humidity	Flashover	To confirm the busbars design (distance between busbars, layers, insulation) and the heat dissipation data shall be included for the HVAC sizing calculation	EPC/OWNER	
Current	Over	Upstream fault, system short circuit	Damage to equipment	Protection coordination study to be conducted during EPC stage and completed prior to commissioning	EPC/OWNER	
Protection and control	No	The power supply of the space heater is down during emergency	Damage to equipment	To design the power supply of the switchgear space heater to be provided from the emergency switchgears for all the switchgears	EPC/OWNER	
Current	Over	Upstream fault, short circuit	Damage to switchgear	The short circuit study, load flow and motor starting study shall be finalised upon completion of data submission from OEM	EPC/OWNER	

(continued)

Table A.5 (continued)

Parameter	Guideword	Cause	Consequences	Recommendation	Action party	Remarks/notes
Protection and control	Inadequate	CT is not sensitive to detect fault	Protection might not pick up the fault	Selection of let through current and relay proposal shall be provided by switchgear VENDOR All functional test for the protection to be conducted during FAT	EPC/OWNER	
Humidity	High	No safety switch and earth fault protection for motor space heater	Condensation on motor could lead to short circuit or motor failure	Space heater supply is provided with RCBO rated 30 mA Safety switch is provided near the motors	EPC/OWNER	NO ISSUE
Protection and control	No	The differential relay between upstream and downstream switchgears not activated	Protection not able to pick up fault	The Class X CT shall be manufactured from the same manufacturer	EPC/OWNER	
Protection and control	No	Absence of earth fault protection for outgoing feeder to external load centres	Trip at upstream	Earth fault protection is provided for the main incomer breaker of all downstream distribution boards	EPC/OWNER	NO ISSUE Provision has been made as below: If more than 100 A, MCCB with LSIG is considered. If less than 100 A, RCB0 is considered Distribution boards located more than 200 m away from the switchgear, earth fault protection shall be provided for the switchgear outgoing feeder also
Interlocking	Availability	Protection cannot be initialised after tripping and during faulty relay	Downstream loads are tripped	To design for latched type lock out relay (86) in an independent relay with bypass switch in case of relay failure	EPC/OWNER	

Table A.6 ESATO—LV switchgear

Elements	Potential hazards and/or operability issues	Causes	Key tasks (safety mitigation plan)	Recommendations	Action by
Switchgear	Fire/explosion— flashover/checklist— visual inspection	Busbar loose connection, hot spots	N/A	To ensure that the arc flash capability is available for all switchgears	EPC/OWNER
Switchgear, transformer	Fire/explosion— flashover/checklist— visual inspection	Inadequate connection, hot spots	To ensure that ample separation/segregation is provided at the termination box based on the number of incomer cables	To ensure that ample separation/segregation is provided at the termination box based on the number of incomer cables. To confirm with OEM for cable termination: 1. No. of cables 2. Size of cables 3. Installation method 4. Temperature/humidity/condensation	EPC/OWNER
Switchgear	Mechanical hazard—blockage	Switchgear cannot be installed or taken out for service or maintenance	To ensure that the switchgear can be installed or taken out for service	To consider the largest panel size for material handling when sizing the substation doors	EPC/OWNER
Transformer switchgear	Electrocution—direct contact	Terminal box too short to install the HV termination kit inside the cable compartment. Additional drop box might be required which will affect the IP and form of the switchgear	To provide adequate space for termination of cables	To confirm the cable compartment size and the gland plate of the cable compartment prior to procurement of switchgear to ensure no drop box is needed. To ensure that vertical support is provided adequately to support the weight of the incoming cables	EPC/OWNER
Switchgear	Electrocution—direct contact	Installing new cables for Phase 2 motors with Phase 1 motors running	To provide adequate space access for installing Phase 2 equipment without interrupting the operation of Phase 1 equipment	To ensure that adequate measures have been taken for phase 2 project cable installation without interrupting current operation	EPC/OWNER

(continued)

Table A.6 (continued)

Elements	Potential hazards and/or operability issues	Causes	Key tasks (safety mitigation plan)	Recommendations	Action by
Switchgear	Accessibility	Future switchgears blocked and not accessible	To provide adequate space access for installing Phase 2 equipment without interrupting the operation of Phase 1 equipment	To provide adequate space for future switchgear. To provide cut out for future expansion panels and cover with metal plate. To ensure that no blocking for operation and maintenance of the Switchgear panels. To ensure that the installation of the Phase 2 Project motor cables shall not be blocked by Phase 1 Project motor cables	EPC/OWNER
Switchgear	OI procedure—PTW	Improper switching	Indicators for switching condition for the circuit breaker	To develop the switching procedure and provide indication at the switchgears	EPC/OWNER
Switchgear	Flashover	Water ingress into switchgears	To prevent water ingress into switchgears	To design the HVAC system such that duct diffusers/blower fans are not installed directly on top of the switchgears	EPC/OWNER
Switchgear	Fire/explosion—flashover/checklist—visual inspection	No tagging at the back	To ensure adequate tagging	Tagging to be provided at both front and back sides of the switchgear	EPC/OWNER
Switchgear	Electrocution—direct contact	PT/VT cannot be accessed during maintenance	To ensure access to PT/VT is available	To design the switchgear with front access to PT/VT with no blockage during maintenance/servicing. To provide all the special tools required to withdraw the PT/VT	EPC/OWNER
Protection relay	OI procedure—work instruction	No indicator of fuse blown for incoming cable measurement system	To provide indication during fault	To design for physical indicator at switchgear panel during CT/PT fault	EPC/OWNER
Bus duct	Fire/explosion—flashover/checklist—visual inspection	Water ingress and dust accumulation at the bus duct cause damage to the bus duct	To ensure bus duct is always clean and dry	To ensure that all penetration for the bus duct are properly sealed with prevention again water and dust ingress	

(continued)

Table A.6 (continued)

Elements	Potential hazards and/or operability issues	Causes	Key tasks (safety mitigation plan)	Recommendations	Action by
Switchgear	Mechanical hazard—others	Switchgear cannot be accessed due to blockage by column	To ensure adequate space around the switchgear for maintenance and operation	To ensure that all switchgears are easily access and is not blocked To provide dummy panels for switchgears panels that is located adjacent to a structural column	

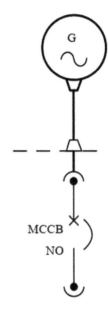

Equipment 0.4 kV 200 kVA.

Fig. A.4 Emergency diesel generator single line diagram

Table A.7 ESSID—emergency diesel generator

Parameter	Guideword	Cause	Consequences	Recommendation	Action party	Remarks/notes
Tagging	Undefined system	Equipment tag is not following PTS standard	Confusion during operation and switching that could lead to operation error	EPC/OWNER to review the tagging system to ensure consistency during EPC stage	EPC/OWNER	NO ISSUE The tagging system is unique for total Project development
Current	Over	Upstream fault, system short circuit	Damage to equipment	Protection coordination study to be conducted during EPC stage and completed prior to commissioning	EPC/OWNER	
Protection and control	No	The power supply of the space heater is down during emergency	Damage to equipment	To design the power supply of the switchgear space heater to be provided from the emergency switchgears for all the switchgears	EPC/OWNER	
Current	Over	Upstream fault, short circuit	Damage to switchgear	The short circuit study, load flow and motor starting study shall be finalised upon completion of data submission from OEM	EPC/OWNER	
Protection and control	Inadequate	CT is not sensitive to detect fault	Protection might not pick up the fault	Selection of let through current and relay proposal shall be provided by switchgear VENDOR All functional test for the protection to be conducted during FAT	EPC/OWNER	
Humidity	High	No safety switch and earth fault protection for motor space heater	Condensation on motor could lead to short circuit or motor failure	Space heater supply is provided with RCBO rated 30 mA Safety switch is provided near the motors	EPC/OWNER	NO ISSUE

(continued)

Table A.7 (continued)

Parameter	Guideword	Cause	Consequences	Recommendation	Action party	Remarks/notes
Protection and control	No	The differential relay between upstream and downstream switchgears not activated	Protection not able to pick up fault	The Class X CT shall be manufactured from the same manufacturer	EPC/OWNER	
Protection and control	No	Absence of earth fault protection for outgoing feeder to external load centres	Trip at upstream	Earth fault protection is provided for the main incomer breaker of all downstream distribution boards	EPC/OWNER	NO ISSUE Provision has been been made as below: If more than 100 A, MCCB with LSIG is considered. If less than 100 A, RCB0 is considered Distribution boards located more than 200 m away from the switchgear, earth fault protection shall be provided for the switchgear outgoing feeder also
Interlocking	Availability	Protection cannot be initialised after tripping and during faulty relay	Downstream loads are tripped	To design for latched type lock out relay (86) in an independent relay with bypass switch in case of relay failure	EPC/OWNER	

Table A.8 ESATO—emergency diesel generator

Elements	Potential hazards and/or operability Issues	Causes	Key tasks (safety mitigation plan)	Recommendations	Action by
Switchgear	Fire/explosion—flashover/checklist—visual inspection	Busbar loose connection, hot spots	N/A	To ensure that the Arc flash capability is available for all switchgears	EPC/OWNER
Switchgear, transformer	Fire/explosion—flashover/checklist—visual inspection	Inadequate connection, hot spots	To ensure that ample separation/segregation is provided at the termination box based on the number of incomer cables	To ensure that ample separation/segregation is provided at the termination box based on the number of incomer cables To confirm with OEM for cable termination: 1. No. of cables 2. Size of cables 3. Installation method 4. Temperature/humidity/condensation	EPC/OWNER
Switchgear	Mechanical hazard—blockage	Switchgear cannot be installed or taken out for service or maintenance	To ensure that the switchgear can be installed or taken out for service	To consider the largest panel size for material handling when sizing the substation doors	EPC/OWNER

(continued)

Table A.8 (continued)

Elements	Potential hazards and/or operability Issues	Causes	Key tasks (safety mitigation plan)	Recommendations	Action by
Transformer switchgear	Electrocution—direct contact	Terminal box too short to install the HV termination kit inside the cable compartment. Additional drop box might be required which will affect the IP and form of the switchgear	To provide adequate space for termination of cables	To confirm the cable compartment size and the gland plate of the cable compartment prior to procurement of switchgear to ensure no drop box is needed. To ensure that vertical support is provided adequately to support the weight of the incoming cables	EPC/OWNER
Switchgear	Electrocution—direct contact	Installing new cables for Phase 2 motors with Phase 1 motors running	To provide adequate access for installing Phase 2 equipment without interrupting the operation of Phase 1 equipment	To ensure that adequate measures have been taken for phase 2 project cable installation without interrupting current operation	EPC/OWNER
Switchgear	Accessibility	Future switchgears blocked and not accessible	To provide adequate access for installing Phase 2 equipment without interrupting the operation of Phase 1 equipment	To provide adequate space for future switchgear. To provide cut out for future expansion panels and cover with metal plate. To ensure that no blocking for operation and maintenance of the switchgear panels. To ensure that the installation of the Phase 2 Project motor cables shall not be blocked by Phase 1 Project motor cables	EPC/OWNER

(continued)

Table A.8 (continued)

Elements	Potential hazards and/or operability Issues	Causes	Key tasks (safety mitigation plan)	Recommendations	Action by
Switchgear	OI procedure—PTW	Improper switching	Indicators for switching condition for the circuit breaker	To develop the switching procedure and provide indication at the switchgears	EPC/OWNER
Switchgear	Flashover	Water ingress into switchgears	To prevent water ingress into switchgears	To design the HVAC system such that duct diffusers/blower fans are not installed directly on top of the switchgears	EPC/OWNER
Switchgear	Fire/explosion—flashover/checklist—visual inspection	No tagging at the back	To ensure adequate tagging	Tagging to be provided at both front and back sides of the switchgear	EPC/OWNER
Switchgear	Electrocution—direct contact	PT/VT cannot be accessed during maintenance	To ensure access to PT/VT is available	To design the switchgear with front access to PT/VT with no blockage during maintenance/servicing To provide all the special tools required to withdraw the PT/VT	EPC/OWNER
Protection relay	OI procedure—work instruction	No indicator of fuse blown for incoming cable measurement system	To provide indication during fault	To design for physical indicator at Switchgear panel during CT/PT fault	EPC/OWNER
Bus duct	Fire/explosion—flashover/checklist—visual inspection	Water ingress and dust accumulation at the bus duct cause damage to the bus duct lePara>	To ensure bus duct is always clean and dry	To ensure that all penetration for the bus duct are properly sealed with prevention again water and dust ingress	

(continued)

Table A.8 (continued)

Elements	Potential hazards and/or operability Issues	Causes	Key tasks (safety mitigation plan)	Recommendations	Action by
Switchgear	Mechanical hazard—others	Switchgear cannot be accessed due to blockage by column	To ensure adequate space around the Switchgear for maintenance and operation	To ensure that all switchgears are easily access and is not blocked. To provide dummy panels for switchgears panels that is located adjacent to a structural column	

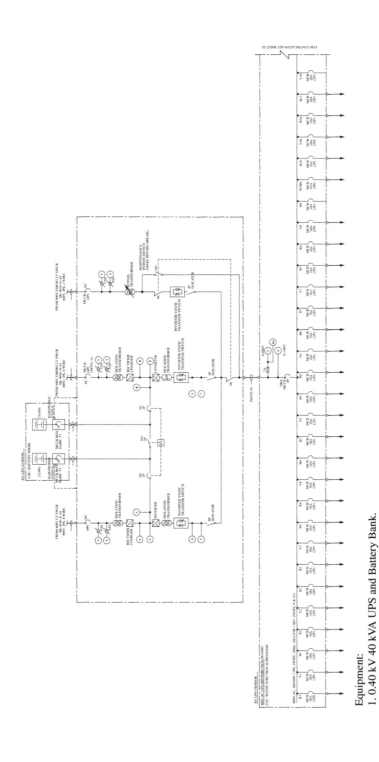

Equipment:
1. 0.40 kV 40 kVA UPS and Battery Bank.
2. 0.40 kV AC UPS Distribution Boards

Fig. A.5 UPS single line diagram

Table A.9 ESSID—UPS

Parameter	Guideword	Cause	Consequences	Recommendation	Action party	Remarks/notes
Tagging	Undefined system	Equipment tag is not following PTS Standard	Confusion during operation and switching that could lead to operation error	EPC/OWNER to review the tagging system to ensure consistency during EPC stage	EPC/OWNER	NO ISSUE The tagging system is unique for total project development
Rating	Inadequate	No automatic changeover facility	Battery no supply during emergency	To clarify with OEM on changeover facility availability of AC UPS To provide operation procedure if auto-changeover facility is not available	EPC/OWNER	
Intertrip	Inadequate	No voting system for F&G	Nuisance tripping	To design the Fire Detection system at the battery room with voting system to avoid nuisance tripping from the detector	EPC/OWNER	

Table A.10 ESATO—UPS

Elements	Potential hazards and/or operability issues	Causes	Key tasks (safety mitigation plan)	Recommendations	Action by	Remarks/notes
Batteries	OI procedure—work instruction	Not enough space to install and access the batteries	To ensure enough space and access to install and maintain the batteries	To adequately size the access door according to dimension of the battery rack for ease of installation and maintenance	EPC/OWNER	
Batteries	OI procedure—work instruction	Access to batteries is difficult for maintenance as the batteries is not maintenance free NiCad battery and needs periodical top up maintenance	To ensure ease of access for battery maintenance	To ensure that the battery rack is designed for ease of operation with 2 tiers To ensure the battery terminals are covered	EPC/OWNER	
Batteries	OI procedure—work instruction	Water dripping or spillage during maintenance of battery electrolytes caused damage to the flooring	To ensure floor is protected against corrosion	To ensure the flooring is corrosion resistance	EPC/OWNER	
Batteries	Toxicity—fumes	There will be hydrogen emission during normal/boost charging after battery maintenance	To ensure there is gas detection system to detect the level of hydrogen in the battery room	To confirm the gas detection system is installed in the battery room	EPC/OWNER	
UPS & batteries	OI procedure—work instruction	UPS and batteries deteriorate due to heat and dust during commissioning	To ensure adequate protection for UPS and batteries during commissioning	To commission HVAC system for UPS and battery room prior to UPS commissioning to ensure that the UPS will be operated in a cool and clean condition		

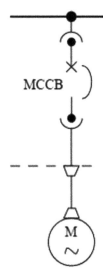

Equipment: 0.4 kV Typical Motor Feeder.

Fig. A.6 Motor feeder

Table A.11 ESSID –0.4 kV Motor feeder

Parameter	Guideword	Cause	Consequences	Recommendation	Action Party	Remarks/Notes
Tagging	Undefined system	Equipment tag is not following PTS Standard	Confusion during operation and switching that could lead to operation error	Review the tagging system to ensure consistency during EPC stage	EPC/OWNER	NB ISSUE The tagging system is unique for total Garraf development
Voltage	Under	Voltage dip in the feeder switch-board	All process motors tripped	To review the capability of sustaining the holding contact within the 0.2 s when voltage dips for motor starters	EPC/OWNER	

Table A.12 ESATO –0.4 kV motor feeder

Elements	Potential hazards and/or operability issues	Causes	Key tasks (safety mitigation plan)	Recommendations	Action by
Motor	OI procedure—LOEA	No LOTO provision	Electrocution—direct contact	To confirm provision of LOTO padlocking facility at the motor LCS/RCU	EPC/OWNER

Lightning Source UK Ltd.
Milton Keynes UK
UKHW020619010822
406667UK00002B/28